MANDELIC ACID

AHA

Madhuri Sharon
Annika Durve
Anuradha Pandey
Manish Pathak

N. Shankaran Nair Research Centre for
Nanotechnology & Bionanotechnology

Jambhul Phata, Ambernath (W), Maharashtra, India

PARTRIDGE

Dr. Madhuri Sharon, is Executive Director of N.S,N, Research Centre for Nanotechnology & Bionanotechnology, she is also Managing Director of MONAD Nanotech Pvt Ltd. She has many patents and publications including four books to her credit.

Print information available on the last page.

To order additional copies of this book, contact
Partridge India
000 800 10062 62
orders.india@partridgepublishing.com

www.partridgepublishing.com/india

CONTENTS

PREFACE

In olden days, distances between townships, limited funds, the lack of available medical professionals **and services all indicated that a girl be not only a wife, mother, and housekeeper, but doctor as well. Folklore healing practices, curative use of herbs, and** different medicinal "family secrets" were stealthily guarded and handed down from one generation to the next. Some of the cures were really valuable and effective whereas some of the grandmother's practices were not actually cures at all, and were just some superstitions and delusion cures which had no practical applications e.g. practice of hanging herbs around a child's neck to assist him lower teeth. Through the years, the superstitions were left behind to start the era of herbal treatment, resulting into birth of Ayurveda in India, Yunani medicine in Middle East and Chinese Herbal treatment and later witch doctors in Europe.

It was while living in tune with nature and studying wildlife that early man learned of the medicinal "powers" of herbs. Animals bitten by a toxic snake survived after chewing snakeroot, a wounded bear rolled in mud to higher heal and escape infection, and old, rheumatoid deer eased their distress and made joints more limber by resting under the therapeutic rays of the sun. In days of old, there was no other solution to treat illness and discomfort, assist heal wounds, or treatment bodily dysfunctions than with pure means.

Although there is no record to establish when plants were first used for medicinal purposes, the use of plants as healing agents is an ancient practice. Over time through emulation of the behavior of fauna, a medicinal knowledge base developed and was passed between generations. As tribal culture specialized specific castes, Vaidyas, Shamans and Apothecaries performed the 'niche occupation' of healing.

By working with, and not against nature, we increase our chance of a more wholesome life, whereas decreasing our threat of illness and untimely bodily limitations and dysfunctions. A wealth of therapeutic resources is there for the taking, if we however open our eyes to the probabilities available to us.

This book is about one such molecule resourced from plant the **Mandelic acid**. This book also emphasizes the use of white biotechnology for synthesis of Mandelic acid. We have prepared this book for the benefit of researchers as well as the industries involved in uses and synthesis of Mandelic acid. Moreover, this molecule is finding many other avenues to express its utility. We hope that this book will be a good support for them.

Madhuri Sharon
Annika Durve
Anuradha Pandey
Manish Pathak

Place: Mumbai
Date: 1st December 2013

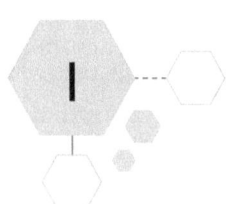

MANDELIC ACID: PHYSICS AND CHEMISTRY

1.1 INTRODUCTION

Chemistry has been known to mankind since time immemorial. Recorded use of knowledge of chemistry goes back to 4000 BC. As early as 3000

Science, has been neither 'discovered' nor 'invented': it is 'revealed'

years ago leather tanning, fruit and grain fermentation and production of copper, iron and steel, brass, silver and gold and their alloys were known to the Vedic Aryans. Moreover, mention of an intoxicant *Somarasa* in *Rig-Veda* is the earliest evidence of fermentation process used in India. Indian steel was highly esteemed in the ancient world because of the heights attained by Indians in metallurgy and engineering. This is evident by the discovery of the almost pure copper stature of Buddha found at Sultanganj and the famous iron Pillar at Mehrauli (Delhi which has been able to withstand rain and weather for centuries without rusting).The Harappan remains of 2500 B.C shows developed metallurgy of copper and bronze. By 1000 BC ancient chemists were extracting metal from their ores,

> *The process of scientific discovery is, in effect, a continual flight from wonder.*
>
> *Albert Einstein*

making pottery and glazes, fermenting beer and wine, making pigments for cosmetics and painting, extracting chemicals from plants for various uses.

Knowledge of Chemistry has fascinated and evolved through the ages and the progress in the direction of revealing the science of Chemistry is continual. The present book is one such effort to understand and use the intricacies of chemistry for production of *Mandelic acid*.

Plants have always been envisaged at the source of useful chemicals. One such plant is Bitter almond or as called in German language *Mandel*. Mandelic acid was discovered while heating amygdalin, (a cyanogenetic glycoside found in many plants including bitter almond, apricot, and wild cherry) an extract of bitter almonds, with diluted hydrochloric acid by Walker and Krieble (1909). Its name is derived from the source plant i.e. *Mandel*. Mandelic acid is an aromatic alpha hydroxyl acid (AHA) with the molecular formula $C_6H_5CH(OH)COOH$. Mandelic acid gained commercial importance with its introduction in cosmetic industries, in the early 1990s, for improving the texture of the skin and reducing the signs of wrinkles.

Figure 1.1 – A–Almond on tree and B-Almonds:
the first known source of Mandelic acid

1.2 STRUCTURE OF MANDELIC ACID

Mandelic acid is chemically known as alpha hydroxyl benzene acetic acid. Mandelic Acid is a phenylglycollic acid, which has a hydroxyl group on the carbon atom next to the acid group. It is an 8-carbon alpha-hydroxy acid with the chemical formula $HOCH (C_6H_5) COOH$. These Alpha Hydroxy acids or 'AHA's are aromatic compounds which occur naturally. Mandelic acid is the smallest aromatic AHA; though it is larger than Glycolic acid, the most popular alpha hydroxy acid. The synonyms used for Mandelic acid are amygdalic acid, uromaline, (S)-alpha-hydroxyphenylacetic acid, 2-phenylglycolic acid, *Acido Mandelico*, alpha hydroxyl benzene Acetic acid.

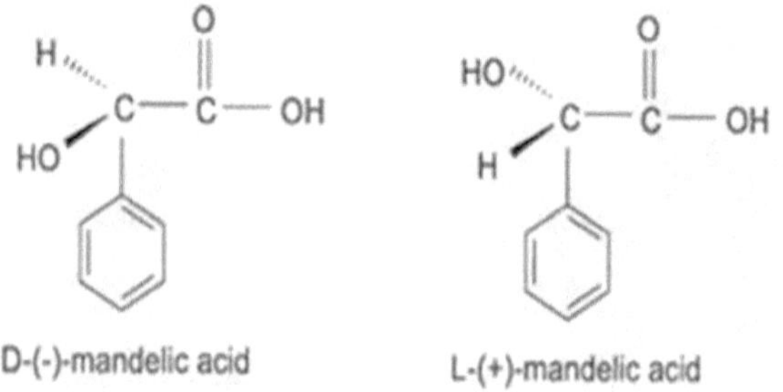

D-(-)-mandelic acid   L-(+)-mandelic acid

Figure 1.2 -Mandelic Acid (Molecular formula: $C_8H_8O_3$)

AHA has dual functionality of acid and alcohol. It has a hydroxyl group on the carbon atom next to the acid group. If the hydroxy group is on the second carbon next to the acid group, it is called beta-hydroxy acid.

It has an asymmetric carbon atom and thus has two chiral isomers; the dextro-, levo-. The D- and L-Mandelic acids are enantiomers (also called enantiomorph; i.e. each molecule is asymmetrical and has the mirror image of the other).

The two enantiomeric forms are likely to affect its pharmaceutical activity differently. Having a pK of 3.37 makes it stronger than glycolic acid, which has a pK of 3.83 at 25°C. Changes in temperature may affect its acidity.

Mandelic acid is an isomer of the cresotinic and the oxymethylbenzoic acids. Since the molecule contains an asymmetric carbon atom, the acid exists in three forms, one being an inactive racemic mixture and the other two being optically active forms and one optically inactive form (para-Mandelic acid). It can be prepared by the action of hydrochloric acid on the benzaldehyde and hydrocyanic acid. The inactive mixture can be converted into active components by fractional crystallization of the cinchonine salt. When the salt of the dextro modification separates first; the ammonium salt may be fermented by *Penicilium glaucum*, where the levo form is destroyed and the dextro form remains untouched; on the other hand, *Saccharomyces ellipsoideus* destroys the dextro form, but does not touch the levo form. A mixture of the two forms in equivalent quantities produces the inactive variety, which is also obtained when either form is heated for few hours to 160 ^{0}C.

Naturally occurring Mandelic acid is found when amygdalin (a cyanogenetic glycoside found in many plants including bitter almond, apricot, and wild cherry) is spirited by hydrolysis with hydrochloric acid, when amygdalin is broken down into glucose, benzaldehyde and prussic acid (hydrogen cyanide) in the presence of sulfuric acid.

$$C_5H_5CHO + HCN + HCl + 2H_2O =$$
$$C_6H_5.CHOH. COOH + NH_4Cl$$

1.3 PROPERTIES OF MANDELIC ACID

Mandelic acid is a water as well as polar organic solvents soluble white crystalline solid. It has a molecular weight of 152.14732 [g/mol] and a Specific Gravity of 1.3 (Water = 1) i.e. its density is 1.3 g/cm^3. It is moderately soluble in water (1g per 6.3 ml) and found to be freely soluble in isopropyl and ethyl alcohol. Melting point of Mandelic acid ranges from 131-135$^{\cdot}$C.

It is a stable compound but is sensitive to light and highly combustible. On exposure to light Mandelic acid becomes dark in colour. Mandelic acid is found to be incompatible with strong bases, strong oxidizing agents, and strong reducing agents. Its structure provides the bacteriostatic property. The general properties of Mandelic acid are summarized below in table 1.1.

Table 1.1: Physical properties of Mandelic acid

Physical Properties	(S)-(+)-Mandelic acid / L-Mandelic acid	(R)-(-)-Mandelic acid/D-Mandelic acid
IUPAC NAME	2- HYDROXY-2-PHENYLACETIC ACID	2- HYDROXY-2-PHENYLACETIC ACID
Appearance	White crystalline powder	White crystalline powder
Structural Formula		
Molecular Formula	$C_8H_8O_3$	$C_8H_8O_3$
Molecular Weight or mass	152.14732	152.14732
Density	1.30 g/cm³	1.30 g/cm³
Specific Rotation	[a]D20 = +154.0 (c=2.8, H_2O)	[a]D20 = -150.0 (c=2.8, H_2O)
Melting point of optically pure	132-135°C	132-135°C
Purity	99%	98%
Solubility in	Chloroform 0.12 M, diethyl ether 0.77 M, methanol 3.54M Water – 1 g/6.3 ml	Chloroform 0.12 M, diethyl ether 0.77 M, methanol 3.54M Water - 1 g/6.3 ml
Stability	Stable, but light sensitive. Combustible. Incompatible with strong bases, strong oxidizing agents, strong reducing agents.	Stable, but light sensitive. Combustible. Incompatible with strong bases, strong oxidizing agents, strong reducing agents.
Acidity (pK$_a$)	3.37	3.37

1.4 CONCLUSION

Mandelic acid is a white crystalline aromatic alpha hydroxyl acid (AHA) with the molecular formula C_6H_5CH (OH) COOH and is soluble in water and polar organic solvents. Due to its chemical and physical properties, it is useful in various industries ranging from medical, cosmetic, antibacterial agent to the chemical industry as an intermediate substance and an analytical reagent. The next chapter deals with the economical importance of Mandelic acid.

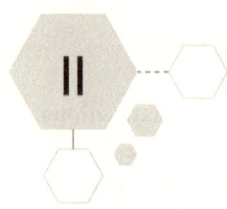

MANDELIC ACID: ECONOMIC IMPORTANCE

2.1 INTRODUCTION: AN ACID THAT HEALS

According to <u>Mandelic acid.org</u>, *"Mandelic acid has long been used in the medical community as an antibacterial, particularly*

> Science never solves a problem without creating ten more.
> *George Bernard Shaw*

in the treatment of urinary tract infections, as well as an oral antibiotic. Lately, Mandelic acid has been receiving greater attention as a topical skin care treatment. Dermatologists now suggest Mandelic acid as an appropriate treatment for a wide variety of skin concerns, from acne to wrinkles; it is especially good in the treatment of adult acne as it addresses both of these concerns. Mandelic acid is also recommended is a pre- and post-laser treatment, reducing the amount and length of irritation caused by laser resurfacing".

In the early 1990s, with the introduction of Alpha Hydroxy acids in cosmetic industries, a skin treatment, which could actually improve the texture of the skin and reduce the signs of wrinkles, became a subject of interest.

Physicians have used it as an internal anti-bacterial for years. Reports indicate that chemically it has a structure similar to some antibiotics. It has been extensively studied to assess its antibacterial action in acne and in preventing bacterial infections after laser skin resurfacing.

Mandelic acid is a useful Chiron for the production of various pharmaceuticals such as semi synthetic penicillin's, cephalosporins and anti-obesity agents. Some of Mandelic acid achievements include:

- Even skin tone and reduced age spots as well as freckles
- Significantly reduced acne breakouts
- Tightening of skin and pores
- Reduced or erased fine lines of wrinkles
- Significantly reduce symptoms of Rosacea
- Help fade signs of melasma

Mandelic acid has a dual role – (i) a potential cosmetic activity and (ii) a well-established antibacterial activity.

Many trials were carried out to determine whether Mandelic acid has the potential for skin ailments. The trials conducted had mainly two aims:

> Work was carried out to see if Mandelic acid had the same effects as that of glycolic acid, which is already an established anti aging agent.
> Second aim was to see if Mandelic acid did have anti-bacterial activity against Gram negative bacteria; as they are known to cause post-surgical infections especially after laser surgery and its usefulness in treatment of acne.

Along with its use in cosmetics, Mandelic acid is also used in herbicides, dye intermediate as well as an analytic reagent

We list below some of the uses of Mandelic acid.

2.2 FOR SKIN CARE

Mandelic acid as an AHA has created a boom in the cosmetic industry. Dermatologists today suggest that incorporation of Mandelic acid into a skin care regimen can help to speed up the results, even when multiple issues are involved simultaneously. Moreover, Mandelic acid is well known for its unique anti-aging activity. It is widely used in skin care products. Its use as a skin care modality was pioneered by Dr James E. Fulton, who developed vitamin A (tretinoin, retin A).

Mandelic acid is extensively used in skin care products. Most AHAs are particularly harsh on skin and can leave the skin feeling raw, red and exposed. Mandelic acid is different from other AHAs in this respect. Because Mandelic acid is a large -form molecule, it is much slower to penetrate the skins layer which allows for a more even treatment. This slow penetration makes it much gentler on the delicate lower dermis and it can be used as a skin care treatment on even delicate skin. Mandelic acids can improve the skin as they can penetrate the outer skin layer and reach the lower dermis. They are able to dissolve the cementing material which holds the dead skin together thus increasing skin turn over and sloughing off dull and rough skin. Mandelic acid gained popularity as an alpha hydroxy acid suitable for all skin types and complexions, giving it a unique advantage. It quickly improves the texture of the skin by diminishing corneocyte cohesion, regenerating cells and promoting a finer, more flexible and better hydrated corneous layer.

Mandelic acid has been studied extensively for its possible uses in treating varied skin problems.

2.2.1 ABNORMAL PIGMENTATION (PHOTO-AGEING) PROBLEM

Abnormal pigmentation includes Lentigenes, Melasma; post inflammatory hyper pigmentation. Mandelic acid is used for rejuvenating photo-aged skin

Lentigenes - Sun-rays are known to cause Lentigenes, which are flat, tan to brown oval spots that are commonly formed due to chronic sun exposure (solar Lentigenes; sometimes called liver spots). They occur most frequently on the face and back of the hands and increase in number with age. Non-solar Lentigenes are associated with systemic disorders such as Peutz-leghers syndrome (in which profuse Lentigenes of the lips occur), multiple Lentigenes syndrome (leopard syndrome) or *Xeroderma pigmentosum*. Lentigenes improve when treated with products containing Mandelic acid.

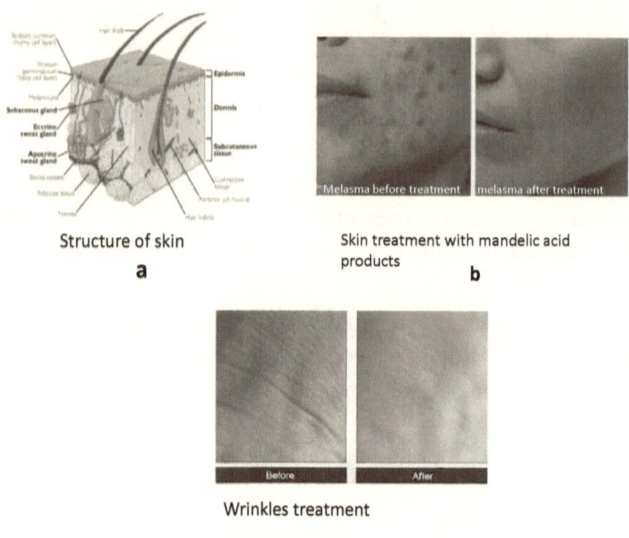

Structure of skin

a

Skin treatment with mandelic acid products

b

Wrinkles treatment

c

Fig 2.1: (a) Schematic diagram of structure of skin and (b) Effect of Mandelic acid in treating Melasma (c) Removal of skin wrinle by Mandelic acid treatment.

Melasma - is a common acquired increase of brownish pigmentation that occurs exclusively in sun-exposed areas. It is particularly common in women, pregnant women and those who are taking oral or patch contraceptives or hormone replacement therapy (HRT) medications. Melasma is thought to be the stimulation of melanocytes or pigment producing cells by the female sex hormones estrogen and progesterone to produce more melanin pigments. This production is enhanced by exposure to sun, during pregnancy, by the use of oral contraceptives and certain anti-epilepsy drugs. The darkening of the skin often seen with melasma was absent when Mandelic acid was used. The effect on dermal and epidermal melasma was different; a sustained, gradual improvement over a period of months is characteristic effect with Mandelic acid products. The use of 10% Mandelic acid lotion has proven to be affective against Melasma. Even though Mandelic acid does show considerable effect in treating skin problems, Combination therapy has shown to yield faster improvement effectiveness and better results. Mandelic acid has the ability to fade dermal Melasma, a condition that is often resistant to topical treatments such as Retin-A or bleaching creams.

Moreover, Mandelic acid also helps to remove skin blemishes at a faster rate. Mandelic acid products used with prescription-strength bleaching agents containing hydroquinone or kojic acid have shown excellent fading results for skin blemishes with no adverse effects thus hastening the benefits of treatment.

2.2.2 ACNE

Mandelic acid has also been used to treat inflammatory noncystic acne, because it possesses anti-seborrheic, keratolytic, and purifying properties, ideal for acne-prone skins. The acid's larger molecule size allows for a slow and even penetration of the acid. Being a natural anti-inflammatory agent, it helps to soothe and calm irritated skin. Acne improvement is remarkable when treated with Mandelic acid in many people having inflammatory pustular, comedonal, and papular acne. Many people have been able to control their acne with Mandelic acid products alone, without the use of traditional acne products. Mandelic acid has been found to be especially useful for treating ladies suffering from both photo-aging and acne. Being an effective acne medication, it has an ability to work without causing any skin irritations. This AHA is also known as super-moisturizers because it improves the skin's ability to trap moisture (not oil).

2.2.3 WRINKLES

Use of Mandelic acid preparations have shown improvement in fine wrinkles and lines in patients with Fitzpatrick skin types I through VI (i.e. almost all the skin types) without any post inflammatory hyper pigmentation. In 1975, Thomas B. Fitzpatrick, MD, PhD, of Harvard medical school, developed a classification system for skin typing. This system was based on a person's complexion responses to suns exposure (Fitzpatrick 1975). Improvement in inflammatory pustular, comedonal and common acne with the use of Mandelic acid is remarkable in many patients.

Though 10% glycolic acid preparations have been used to remove fine wrinkle lines and photo-aged skin; but this use of glycolic acid (at higher concentrations of 30 % to 70%) has lead to certain problems like skin irritation and erythema. Moreover, it's a time consuming procedure taking almost months to years altogether. Whereas no side effects were seen when Mandelic acid was used and improvement could be seen in fine wrinkles and lines in patients with Fitzpatrick skin types, without any post inflammatory hyper pigmentation.

Mandelic acid also addresses the other main causes of wrinkles like hyper-pigmentation by accelerating the cellular turnover rate, ridding the skin of dead skin cells that dull the complexion and enhancing the natural creases in the skin.

2.2.4 BREAKDOWN OF COLLAGEN

Mandelic acid addresses the breakdown of collagen-one of the main cause of wrinkles. Collagen is an essential fiber in skin that replenishes itself easily when the skin is young. However, as the skin ages, collagen production slows down and the fibers weaken, leading to wrinkle formation. Mandelic acid helps to strengthen the collagen in skin, thus delaying and in some cases reversing the signs of aging and reducing the appearance of wrinkles in the skin

Usage of Mandelic acid with concentrations between 10% to 20% also shows a marked increase in glycosaminoglycan (precursors to collagen) levels in the skin. The ability of Mandelic acid to stimulate collagen production, results in the softening of fine lines and wrinkles.

Mandelic acid has a great capacity for shaping the skin. It has a pka of 3.37 due to which its capacity to increase the synthesis of collagen, elastin and glycosaminoglycans in the dermis is greater than other AHAs such as glycolic acid (pka 3.83).

2.2.5 LASER SKIN RESURFACING AND MANDELIC ACID

Mandelic acid has been found to be effective on sensitive skin of all types, even on rosacea-prone skin; during preparation of the skin for laser peeling and in helping the skin heal after laser surgery. Some acne rosacea patients that were treated with Mandelic acid also showed improvement. Rosacea is a chronic inflammatory disorder characterized by facial flushing, erythema, papules, pustules and in severe cases rhinophyma. It most commonly affects patients aged 30-50 with fair complexions.

Mandelic acid has also been added in an algae extract gel or lotion base for topical use. These preparations also contain certain combinations of topical vitamins including vitamins A, C, E, and D3 and sunscreen with SPF 15.

2.2.6 AS AN EXFOLIATING AGENT:

Mandelic acid acts by weakening the bonds that hold dead skin cells together. This process allows dead skin cells to exfoliate off the skin more quickly and creates a denser, tighter skin cell composition. Thus it is used as a strong exfoliating agent. Mandelic acid has proved itself stronger than glycolic acid. The gentle exfoliating action of Mandelic acid has proven to be non-irritating, which is important to acne prone skin. Mandelic acid products also prevent irritation of skin and erythema production that often accompanies skin treatments with glycolic acid in 30% to 70% preparations used for peeling. The exfoliating action does result in more even toned and a brighter glow of the skin. The increased shedding of dead skin cells on the outermost layer of the skin helps in the penetration of other skin care ingredients, allowing them to penetrate more easily and perform more effectively. The AHA properties of Mandelic acid help to remove dead skin that clog the pores, allowing them to be washed away before they cause a problem. It exfoliates the epidermis much better than with salicylic acid, which cannot penetrate very deeply into the epidermal layers. Thus, making the skin smoother, tighter and more even in tone. The combined activity of AHA and the lipid permeability aids in the appearance of smaller pores. Salicylic-Mandelic acid combination peels (SMPs) are a newer product, and have been found to be more effective than using salicylic acid alone. Repeated chemical peeling using Mandelic acid was found to be useful in treating acne, melasma (brown spots), Lentigenes (large freckles) and fine photo-aging damages to the skin causing wrinkles and dullness in skin texture.

2.3 MANDELIC ACID AS AN ANTI-MICROBIAL AND ANTI-SEPTIC AGENT

Mandelic acid is well known for its anti-microbial and antibacterial activities. Its structure provides the bacteriostatic property. Its antibacterial activity is seen against both Gram-positive and Gram-negative bacteria. Many acne patients are given antibiotics both systemically and topically. Some of these patients have shown resistance to treatment using antibiotics but have shown considerable response to the Mandelic acid. This proves that Mandelic acid does have an antibiotic nature. Patients with Gram-negative folliculitis also showed improvement while using Mandelic acid products. Many acne patients who are resistant to antibiotics, given both systemically and topically, have responded positively to Mandelic acid

Mandelic acid also has sebo-regulating properties. The use of Mandelic acid products has shown remarkable improvement in treating adult female patients suffering from both photo-aging and acne.

2.3.1 AS URINARY TRACT ANTISEPTIC

When the pH of the urine is below 5.5, it has antibacterial activity; hence substances that can acidify urine are used to treat urinary tract infections. Since Mandelic acid is anti-microbial; and can acidify urine, it has been used in medicine for many years as a urinary antiseptic in the form of Methenamine mandelate. Mandelamine is available for oral use as film-coated tablets. This combination has the urinary antiseptic action of both methenamine and Mandelic acid. It is found to inhibit *Staphylococcus aureus, Bacillus, Proteus, Escherichia coli,* and *Aerobacter aerogenes*, when the urine has 35g to 50g/100L concentrations methenamine mandelate. Methenamine is absorbed but remains inactive until it is excreted by the kidney and concentrated in the urine. Acid urine is required for antibacterial action, with maximum efficacy occurring at a low pH of 5.5 or less. Mechanism of action is such that in acid urine, Mandelic acid exerts its antibacterial action and also contributes to the increase in acid content of the urine. Mandelic acid is excreted by the glomerular filtration and tubular excretion. As mentioned above, chemically, Mandelic acid has a structure similar to that of other well-known antibiotics. It is a nontoxic substance that even after being ingested orally is excreted in the urine. A 1% solution of Mandelic acid having pH 2.4 is used as a bladder irrigation fluid in hospitals to prevent urinary tract infections associated with indwelling urethral catheterization. It is also used as an oral antibiotic.

2.4 VAGINAL CONTRACEPTIVE:

Presently marketed vaginal barrier methods are cytotoxic and damaging to the vaginal epithelium and natural vaginal flora when used frequently. Novel non-cytotoxic agents are needed to protect men and women from sexually transmitted diseases. According to the results of in vivo animal experiments and in vitro studies conducted by Dr. Lourens J. D. Zaneveld from Rush-Presbyterian-St. Luke's Medical Center, Chicago; showed that Mandelic acid condensation polymer (SAMMA) when used as a vaginal contraceptive inhibits sperm activity, and seems to have a broad antimicrobial activity. It is non-mutagenic, non-toxic and nonirritant in action. They have studied the efficacy and safety of SAMMA in inhibiting sperm function by inhibiting hyaluronidase and acrosin activity. SAMMA also caused sperm acrosomal loss. It was found to be effective in preventing the transmission of sexually transmitted disease. According to their report in the November issue of Fertility and Sterility, SAMMA prevented infection by Human Immunodeficiency Virus -1 and Herpes Simplex Viruses 1 and 2. SAMMA is highly effective against the CCR5 and CXCR4 isolates of HIV tested in primary human macrophages and peripheral blood mononuclear cells. SAMMA was found to inhibit infection of cervical epithelial cells by HSV. SAMMA, although not a sulfonated or sulfated polymer, blocks the binding of HIV and HSV to cells by targeting the envelope glycoproteins gp120 and gB-2, respectively, and also inhibits HSV entry postattachment. This exciting new, noncytotoxic vaginal microbicide candidate also prevented *Chlamydia trachomatis* infection and could inhibit the multiplication of *Neisseria gonorrhoeae*. A US patent on the antimicrobial and contraceptive properties of SAMMA has been granted (Fertil Steril 2002; 78:1107-1115).

2.5 OTHER USES OF MANDELIC ACID

Mandelic acid has been used as a chiral synthon for the production of various pharmaceutical products such as semi synthetic penicillins, cephalosporins.

It has played a role in reducing the weight in products like anti-obesity agents.

Its killing effect is useful as it is also used in various herbicide formulations.

Mandelic acid is also used as a dye intermediate as well as an analytic reagent in various reactions.

Enantiomerically pure Mandelic acid is used as chiral impregnating reagents and as mobile phase additive for direct resolution of the enantiomers of the racemic drugs ketamine and lisinopril by TLC method.

Styrene is a yellowish oily liquid widely used in the production of various plastics, polyester resins and synthetic rubber and has a toxic effect on humans as a carcinogen. The major metabolite of styrene is found by hepatic cytochrome P-450 to be present in two enantiomeric forms, R-(+)-styrene-7,8-epoxide and S-(-)-styrene-7,8-epoxide (STO) this product is further metabolised to form chiral R-and S- 1-hydroxy-1-phenyl-acetic acid (R-and S-Mandelic acid, MA). Metabolites formed are excreted in urine, with 85% of the absorbed styrene eliminated as Mandelic acid. Thus, the concentration of Mandelic acid particularly its enantiomers, in urine has been used as a biological indicator of styrene exposure in workers.

Blood coagulation is an important process involved in both haemostasis i.e. the prevention of blood loss from a damaged vessel, and thrombosis i.e. the formation of a blood clot in a blood vessel. Coagulation results due to a complex series of enzymatic reactions. One of the ultimate steps in this series of reactions is the conversion of the proenzyme prothrombin to the active enzyme thrombin. Thrombin is known to activate platelets, leading to platelet aggregation, converts fibrinogen into fibrin monomers, which polymerizes spontaneously into fibrin polymers and activates factor XIII, which in turn crosslinks the polymers to form insoluble fibrin.

By inhibiting the aggregation of platelets and the formation and crosslinking of fibrin, effective inhibitors of thrombin would be expected to exhibit antithrombotic activity. There are various compounds and derivatives that are useful as competitive inhibitors of trypsin-like proteases, such as thrombin and thus, in particular, in the treatment of conditions where inhibition of thrombin is required (e.g., thrombosis) or as anticoagulants. One such compound is D,L-alpha-phenethylamine (PEA) which is used by dissolving it together with D (-) Mandelic acid in an aqueous solution comprising a second acid other than Mandelic acid. D, L-phenylalanine is resolved with optically active Mandelic acid in the presence of a stabilizing acid. The L-Mandelic acid used to recover the L-isomer desired and does not provide for recovery of the resolving agent.

2.6 THE DISADVANTAGES OF USING MANDELIC ACID IN HEALTH CARE.

Mandelic acid is found to be hazardous in the case of ingestion and inhalation. It can act as an irritant. Its toxicology is not fully investigated. Though some material safety data sheets (MSDS) consider it to be non hazardous for air, sea and road freight (as per the directive 67/548/EEC).

2.7 CONCLUSION

Years of trials and tests have made us reach a conclusion that Mandelic acid products, used alone or in tandem with antioxidant vitamins, have multiple beneficial effects on skin treatment-including antibacterial effects and improvement in photo-aged skin, acne, abnormal pigmentation, and skin texture. It was also found to be safe to be used in darkly pigmented skin types, when the Mandelic acid products were compared with glycolic acid and tretinoin. The market for Mandelic acid products shows that the use of Mandelic acid in combination with other acids like 10 % malic acid, jojoba seed (skin scrub), mint leaves and certain algal products has been found to be more effective. Mandelic acid is also combined with a sun screen like SPF 15 to act as a sun screen lotion.

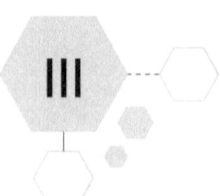

III

MANDELIC ACID: CHEMICAL SYNTHESIS

3.1 INTRODUCTION

Mandelic acid was discovered while heating amygdalin, an extract of bitter almonds, with diluted hydrochloric acid. With the use of Mandelic acid at a large scale in the cosmetic and pharmaceutical industry, the extraction of the acid from bitter almonds tends to

> *Men love to wonder, and that is the seed of science.*
>   *Ralph Waldo Emerson*

be tedious. Along with the lengthy purification processes, the amount obtained in the end is normally very less. To overcome these problems, chemical synthesis of Mandelic acid is carried out. The process is cheaper and results in large amount of Mandelic acid being produced. It is usually prepared by the acid-catalyzed hydrolysis of mandelonitrile, which is a cyanohydrin of benzaldehyde. Mandelonitrile can also be prepared by reacting Benzaldehyde with sodium bisulfite to give the corresponding adducts, forming mandelonitrile with sodium cyanide, which is hydrolyzed to form Mandelic acid. Mandelic acid can also be prepared by base hydrolysis of phenylchloroacetic acid and dibromacetophenone. It is formed by heating benzoyl formaldehyde with alkalis.

3.2 PRODUCTION OF MANDELIC ACID FROM AMYGDALIN

As mentioned earlier, Mandelic Acid was discovered by Walker and Krieble (1909) while heating amygdalin extracted from bitter almonds with diluted hydrochloric acid. **Amygdalin** (Greek word for almond is *amygdálē*), $C_{20}H_{27}NO_{11}$, is a glycoside. Amygdalin is extracted from almond by boiling in ethanol; on evaporation of the solution and the addition of diethyl ether, amygdalin is precipitated as white minute crystals. Hydrochloric acid decomposes Amygdalin into Mandelic acid, D-glucose, and ammonia. Naturally occurring amygdalin has the *R* configuration at the chiral benzyl center. Under mild basic conditions, this stereogenic center epimerizes, producing **Neoamygdalin.**

3.3 PRODUCTION OF MANDELIC ACID FROM BENZALDEHYDE

Mandelic acid is chemically synthesized from benzaldehyde for large-scale production. Benzaldehyde is commonly employed to confer almond flavor.

Synthesis of Mandelic acid using Benzaldehyde and potassium cyanide - Mandelic acid is prepared from benzaldehyde through the cyanohydrin. This cyanohydrin is produced by interaction between benzaldehyde bisulfite and potassium cyanide. This reaction needs vigorous stirring until the oily aldehyde is all converted into the crystalline bisulfite addition compound. To this addition mixture, after cooling at room temperature; potassium cyanide and water are added and stirred to ensure that all the solids have dissolved. This interchange reaction eliminates the hazard of working with volatile, toxic hydrogen cyanide, but still potassium cyanide is dangerous if safety measures are not observed properly. This interchange reaction is reversible and the excess cyanide is used to shift the equilibrium in favor of the cyanohydrin. Mandelonitrile; a liquid cyanohydrin is formed. Mandelonitrile separates out in the form of thick oil; it is further processed without delay because Mandelonitrile undergoes changes if kept standing. Mandelonitrile is extracted with ether and the solution is carefully washed free of cyanide ion. The mixture is then rinsed with small amounts of ether and water. This allows to get rid of all the cyanide. Ether is recovered for reuse.

Mandelonitrile is used as the substrate for the production of Mandelic acid. It is a yellow oily liquid, insoluble in water but soluble in alcohol, ether, $CHCl_3$. It is stable in anhydrous conditions at room temperature. Mandelonitrile has a boiling point of 170° C and a freezing /melting point of -10° C. Mandelonitrile is chemically stable; but it is unstable in water even at room temperature and decomposes. During first half hour decomposition rate is very slow then it is substantially fast and gets decomposed completely in four to six hours. After decomposition it may produce hazardous products like hydrogen cyanide, nitrogen oxides, carbon monoxides, carbon dioxide, etc. Mandelonitrile is an intermediate compound formed during the chemical synthesis of Mandelic acid from benzaldehyde.

Mandelonitrile thus formed is then hydrolyzed with hydrochloric acid. The solution is added into a 125 ml distilling flask containing 15 ml each of concentrated hydrochloric acid and water. The entire mixture is mounted on a steam bath for distillation to be carried out. Continuous heating on the steam bath with frequent swirling allows the proper mixing of the layers and promotes hydrolysis. It is seen that the initially lighter-than-water layer of mandelonitrile gradually changes to an oil of density greater than that of the aqueous acid. The heating is continued for about one and half hour to complete the reaction and then the solution obtained is cooled at room temperature.

An intermediate is formed which is ketimine hydrochloride, $(C_6H_5CH(OH)C=NH*HCl)$. Ketimine hydrochloride is extracted using ether in a separating funnel. This etheral extract is mixed with benzene and distilled to remove water. Benzene helps in the displacement of ether and helps in crystallization as Mandelic acid is much less soluble in this solvent than in ether. Water is eliminated gradually by azeotropic distillation and the boiling point rises as ether and water are eliminated. The distillation process is continued till the solution becomes clear and there are no traces of water in the flask. The flask is removed to check if any traces of ammonium chloride are obtained either in the form of a gum or solid material. The crystallization of Mandelic acid soon begins to give white crystals of Mandelic acid.

The crystals of Mandelic acid are allowed to stand in contact with mother liquor for several days. This results in the needles gradually changing to a granular, sugar like crystals having one molecule each of Mandelic acid and benzene. In benzene, Mandelic acid remains stable. When exposed to air at a temperature below 32.6°C, the benzene of crystallization evaporates and Mandelic acid is left behind in the form of a white powder.

This reaction results in the formation of R-Mandelic acid and S- Mandelic acid. A racemization reaction takes place, which converts the S-Mandelic acid to R-Mandelic acid.

Synthesis of Mandelic acid using Benzaldehyde and Chloroform -Mandelic acid can also be synthesized from benzaldehyde with chloroform. In this methods ultrasonic irradiation is applied where tri-ethyl benzyl ammonium chloride (TEBA) and polyethylene glycol-800 (PEG-800) as a complex phase transfer catalyst are used. The reaction is carried out at 60 °C for 2 hrs only. This process is very advantageous as its reaction time is very short and the yield is comparatively higher than the classical method.

Diasteromeric crystallization with L-phenylalanine as the resolving agent is used for the chiral resolution of Mandelic acid. The carboxyl and amine groups of L-phenylalanine tend to form hydrogen-bond with the hydroxyl and carboxyl groups of Mandelic acid for the preferential formation of a complex of L-Mandelic acid-L-phenylalanine due to the stereo-configurations of the functional groups, and Mandelic acid crystallizes out as a diastereomeric salt. This process results in the enrichment of L-Mandelic acid in the form of diastereomeric crystals.

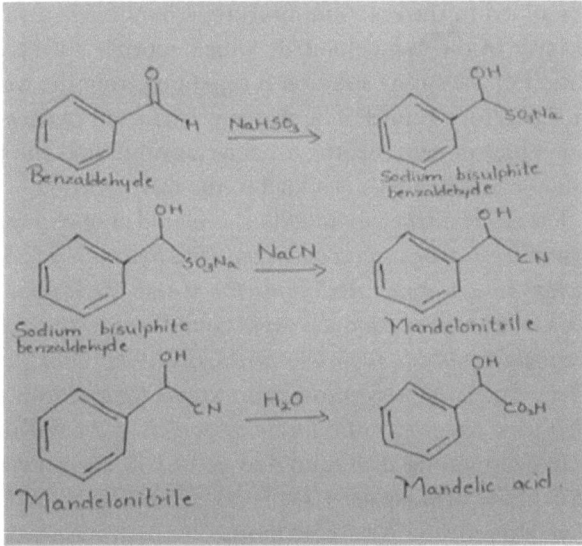

**Fig 3.1: Chemical process for Mandelic
acid synthesis from Benzaldehyde**

Detailed Protocol for Mandelic Acid synthesis

In a 4-L wide mouthed glass jar' fitted with a mechanical stirrer, is placed a solution of 150g. (3 moles) of sodium cyanide (note: - *the reaction and the subsequent hydrolysis are carried out in a good hood as some hydrogen cyanide is liberated. The sodium cyanide used is the technical "cyan-egg" containing about 92-95 % of cyanide.*) In 500 cc of water and 318g (3 moles) of u.s.p. benzaldehyde. The stirrer is started and 850 cc of a saturated solution of sodium bisulphate (note:- *this saturated solution is best prepared by stirring 1500 g of technical sodium bisulphate (97-100%) with 2 l of water and filtering to remove the excess salt. The specific gravity of this solution is 1.37-1.39.)* is added to the mixture, slowly at first and then in a thin stream. The time of addition is ten to fifteen minutes. During the addition of the first half of this solution, 900g of cracked ice is added to the reaction mixture, a handful at a time. The layer of the mandelonitrile which appears during the addition of the sulfite solution is separated from the water in a separatory funnel. The water is extracted once with about 150cc of benzene, the benzene is evaporated and the residual mandelonitrile is added to the main portion.

The crude nitrile (about 290cc) is placed at once (note: - *the mandelonitrile should be mixed with hydrochloric acid as soon as it is separated from the water. This appears necessary in order to avoid a rapid conversion to the acetal of benzaldehyde and mandelonitrile, C_6H_5CH [OCH(CN) $C_6H_5]_2$. Hence if it is allowed to stand long before the hydrolysis, the yield of Mandelic acid is reduced*) in a 25 cm evaporating dish, and 425 cc of c.p. concentrated hydrochloric acid (sp.gr. 1.19) is added. The hydrolysis is allowed to proceed in the cold (note: - *the hydrolysis can be carried out in the hot, but the final product may be deeply colored.*) for about 12 hours, after which the mixture is heated on a steam bath to remove the water and excess hydrochloric acid. After heating for five to six hours it is advisable to cool the mixture (note:- *it is advisable to stir the mixture during the cooling in order to break up the lumps and thus obtain a product that can be more easily filtered.*) and filter the ammonium chloride and Mandelic acid mixture that separates. The filtrate is then evaporated to dryness. This residue is added to the solid material obtained before. The product is deeply colored and must be dried in the air and light for at least 24 hours. The total yield of crude Mandelic acid – ammonium chloride mixture is 370-390g, depending on the amount of moisture. The mixture of ammonium chloride and Mandelic acid is ground in a mortar, transferred to a 2-L flask, and washed twice with 750 cc portions of cold benzene (*If the crude product is not first washed with cold benzene the final product is usually colored. Very little Mandelic acid is lost* *by this washing*) the insoluble portion is transferred to a suction funnel and sucked dry.

Extraction of Mandelic acid – there are two methods to extract the Mandelic acid from the ammonium chloride

(A) <u>Extraction with Benzene</u>- Inorder to separate Mandelic acid from ammonium chloride extraction with hot benzene is carried out. This best done by dividing the solid mixture into ten approximately equal parts (note: - *the entire amount of the ammonium chloride-Mandelic acid mixture may be boiled with the benzene, but this gives a super saturated solution of the acid in the benzene and much difficulty is met in the filtration)*. The solubility of Mandelic acid in hot benzene is approximately 1 g in 50 cc. Extraction is carried out in a Soxhlet apparatus.

One of these portions is placed in a flask with boiling benzene. After a few minutes the hot benzene solution is decanted through a previously heated suction funnel having large holes so as not to be clogged by the Mandelic acid, which begins to crystallize as soon as the solution cools slightly. Very slight suction is applied during filtration. The filtrate obtained is then allowed to cool in an ice bath and the crystallised Mandelic acid is filtered with suction. The benzene is returned to the extraction flask containing the residue from the first extraction, and the new portion of the ammonium chloride-Mandelic acid mixture is added and extracted as before. This process is continuously repeated until Mandelic acid is completely obtained and ammonium chloride residues reemain. Usually two or three extractions of the ammonium chloride residues after the addition of the last portion of the crude mixture are necessary in order to obtain all the Mandelic acid. The yield of pure white Mandelic acid melting at 118^0 C is 229-235 g

Extraction with ether - is faster, especially when several runs are to be made. Moreover, the yield by this method is the same as by benzene extraction which is better for small preparations, or when a single run is to be made.

For ether extraction the solid mixture is transferred to 2-L flask and shaken for ten minutes with 50 cc of ether. The ether solution is decanted through the suction filter and the solid thrown onto the filter and pressed dry. The solid is then added back to the flask and shaken with 400 cc of ether. The mixture obtained is filtered by suction and the solid is washed twice on the filter with 250 cc parts of the ether. Every portion is drained through filter several times. and the filtrate obtained is filtered through an ordinary funnel. The ether solution is placed in round bottomed flask and 750 cc of toluene or xylene instead of benzene is added. The mixture is distilled on a steam bath through an efficient fractioning column as long as it distils easily and about 1100-1400 cc of distillate is collected. when the temperature of vapor rises to 70˚C. The mixture is heated over a free flame until the temperature in the column reaches 95°; about 300 cc distils. A few porous chips are added to prevent bumping. As Mandelic acid is apt to separate the ether-toluene solution should not stand long before distillation. The heating with steam and free flame should be done quickly as prolonged heating lowers the yield. The hot residual liquid is poured into a large beaker immersed in ice water. The liquid (about 900 cc) is stirred till it becomes a thick crystal mush. The cooling is continued for two hours with occasional stirring to 5-10° C.The mixture is filtered by suction and the solid is pressed to dry, followed by thorough washing on the filter with 300 cc of toluene in several parts.

Flow Chart 3.1 – Production of Mandelic acid using chemical methods

150 g of sodium cyanide (3 moles) + 318 g of Benzaldehyde (3 moles) are added to 500 cc of water.

 Solution is stirred.

850 cc of Saturated Sodium bisulphate (slowly)

 900 g of crushed ice is added

Layer of Mandelonitrile appears is separated from water using a separating funnel. The water is extracted once with 150cc of benzene, the benzene is evaporated and the residual mandelonitrile is added to the main portion

Crude mandelonitrile obtained (290cc) is taken in an evaporating dish, 425 cc of Conc. HCl is added.

 Hydrolyzed in Cold conditions, 12 hrs.

The mixture is heated for 5-6 hrs on a steam bath to remove water and excess of hydrochloric acid. Cooled.

Ammonium chloride and Mandelic acid mixture that separates out is filtered & evaporated to complete dryness

 Mixture obtained is air dried for 24 hrs

The total yield of crude Mandelic acid – ammonium chloride mixture is 370-390g

Ammonium chloride and Mandelic acid mixture is ground in a mortar, transferred to a 2-1 flask & washed twice with 750 cc cold benzene

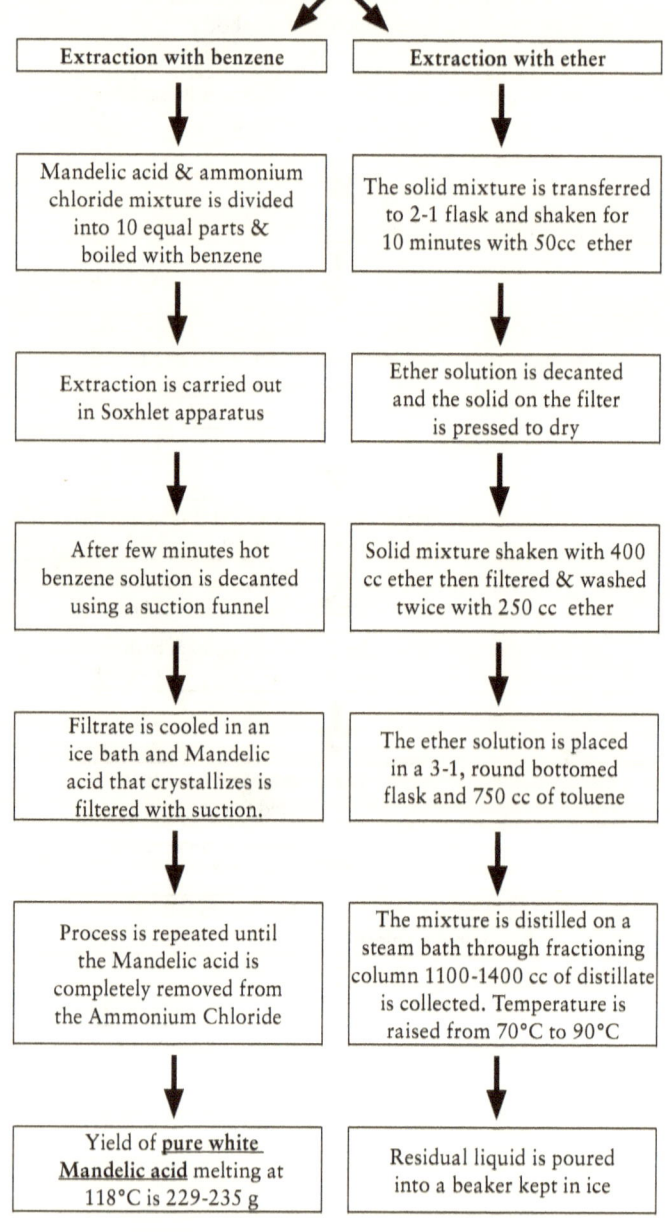

Extraction with benzene	Extraction with ether
Mandelic acid & ammonium chloride mixture is divided into 10 equal parts & boiled with benzene	The solid mixture is transferred to 2-1 flask and shaken for 10 minutes with 50cc ether
Extraction is carried out in Soxhlet apparatus	Ether solution is decanted and the solid on the filter is pressed to dry
After few minutes hot benzene solution is decanted using a suction funnel	Solid mixture shaken with 400 cc ether then filtered & washed twice with 250 cc ether
Filtrate is cooled in an ice bath and Mandelic acid that crystallizes is filtered with suction.	The ether solution is placed in a 3-1, round bottomed flask and 750 cc of toluene
Process is repeated until the Mandelic acid is completely removed from the Ammonium Chloride	The mixture is distilled on a steam bath through fractioning column 1100-1400 cc of distillate is collected. Temperature is raised from 70°C to 90°C
Yield of **pure white Mandelic acid** melting at 118°C is 229-235 g	Residual liquid is poured into a beaker kept in ice

> About 900 cc liquid is
> stirred till thick & cooled,
> pressed to dryness & washed
> with 300cc toluene

3.4 DERIVATIVES OF MANDELIC ACID

With the use of Mandelic acid in various cosmetic and pharmaceutical industries, various combinations of Mandelic acid have been found to be quite useful. Listed below are some methods to produce various derived forms of Mandelic acid.

3.4.1. ACETYL-MANDELIC ACID

Mandelic acid and acetyl chloride are mixed in a Claisen distilling flask. The reaction is carried out without the application of heat. Occasionally the application of a little heat is necessary to bring about a more rapid acetylation. As soon as a clear solution is obtained, the flask is warmed in a water bath and the acetyl chloride is distilled off. The remaining acetyl chloride is removed by prolonged drying in vacuum. The acetyl Mandelic acid obtained is crystallized in large, round, white clusters after 1-2 days of standing.

Crude extract of acetyl-Mandelic acid still contains small amount of acetyl chloride is taken. To this solution thionyl chloride is added. The reaction begins without warming but it is essential to reflux for four hours to complete the reaction. Excess refluxing of the acetyl-Mandelic acid with the thionyl chloride tends to lower the yield. The excess thionyl chloride is distilled off and the residue obtained is distilled under reduced pressure. The pressure is kept as low as possible, to avoid the formation of tar during the distillation. A colorless liquid is obtained.

3.4.2. [1-14C] - MANDELIC ACID

[1-14C]- Mandelic acid is used as radiotracer for the absorption, distribution, metabolism and excretion studies. Recently microwave technology has taken an undeniable place in chemical laboratory practice as a very effective and non-polluting method of activating reactions. Microwaves provide fast heating of the chemicals at or above their boiling points thus enhancing the reaction rates and dramatically reducing the reaction times in comparison with conventional heating. The combination of solvent free reaction conditions and microwave irradiation leads to substantial reduction in reaction times and enhancement of yields, with several advantages of an eco-friendly approach, termed "green chemistry". Microwave irradiation has been efficiently employed for the radio synthesis of racemic [1-14C]-Mandelic acid starting from benzyl bromide. Benzyl bromide on cyanation with $K_{14}CN$ followed by acid hydrolysis in a microwave oven gave [1-14C] phenylacetic acid, which on a- bromination reaction followed by hydrolysis under microwave irradiation for 2 minutes, furnishes the [1-14C] - Mandelic acid.

3.4.3. ([+-])-(RING [SUP 13]C[SUB 6])-MANDELIC ACID

A one-pot synthesis of (ring [sup 13]C[sub 6])-Mandelic acid has been reported in some places. [Ring [sup 13]C[sub 6]]-benzaldehyde is cyanosilylated along with trimethylsilyl cyanide (TMSCN)/ZnI[sub 2]. The resulting cyanosilylated adduct is further hydrolyzed with concentrated hydrochloric acid. This process is carried out without further purification. The process involves the evaporation to dryness. The extraction of the (ring [sup 13] C[sub 6])-Mandelic acid is carried out using hot benzene. After crystallization the purity of the product is checked using HPLC with UV detection

3.4.4 2-AND 4- HYDROXYMANDELIC ACID

Hydroxy Mandelic acids are very valuable as they are versatile starting materials for the production of compounds that are used in fine chemicals for production of therapeutic as well as pharmaceutical chemistry and agricultural chemistry. 4-hydroxyMandelic acid is the intermediate product for the preparation of 4-hydroxybenzaldehyde used in synthesis of fine chemicals.

2- Hydroxy Mandelic acid is an intermediate product for the preparation of agrochemical compound ethylenediamine-N, N'-bis(2-hydroxyphenyl acetic acid) which is used in pharmaceutical compounds.

It is prepared by condensing glyoxylic acid with phenol; the formed reaction mixture results into formation of 2-and 4-hydroxyMandelic acid and remnants of phenol; that are eluted separately in a column comprising of an anion exchange resin. First the excess phenol is eluted out followed by the separation of the 4-hydroxyMandelic acid and finally 2-hydroxyMandelic acid, both in the form of acid and salt, depending on the eluent used.

3.5 DRAWBACKS OF CHEMICAL SYNTHESIS OF MANDELIC ACID

Even though there have been known physical and chemical methods for the production of Mandelic acid, such as optical resolution by fractional crystallization, optical resolution by chromatography, or stereo-selectively synthesizing methods in organic chemistry, these methods have certain drawbacks. First and foremost is the use of chemical solvents like benzaldehyde and potassium/sodium cyanide. Benzaldehyde causes eye irritation and causes allergic reactions. It is considered harmful when ingested or inhaled. There is limited evidence that this chemical may act as a carcinogen in laboratory animals. At high concentration, it acts like a narcotic. Sodium or potassium cyanide cause irritation of the eye and skin when in contact. It has the ability to permeate the skin, and can produce inflammation and blistering. Severe over-exposure can produce lung damage, choking, unconsciousness or death. This chemical when in contact with any acid is converted into hydrogen cyanide. These cyanides are released during the process thus making them pollutants and not eco friendly. Other than that the operation is complicated or the yield and optical purity is low.

3.6 CONCLUSIONS

This chapter has dealt with various production methods of Mandelic acid, which involves use of hazardous chemicals. Moreover, productions of various Mandelic acid derivatives have also been discussed.

In the end of the chapter attention is focussed on hazards of the chemical synthesis method. Many of the methods used and produced as by-product have been proven toxic to humans as well as the eco system. For these reasons, attention is being directed to the production of Mandelic acid by eco friendly methods. These methods use the process of White Biotechnology (plant and microorganism enzymes) and green chemistry (eco friendly synthesis) for the production of Mandelic acid. The later chapters are dedicated to the biosynthetic mechanism and role of enzymes in production of Mandelic acid.

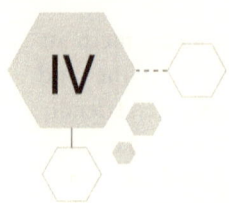

IV

ENZYMES AND BIOTRANSFORMATIONS

4.1 INTRODUCTION

The living cell is the site of incredible biochemical activity called metabolism involving changes which are chemical and physical, which takes place continually for life sustenance.

Technology does not drive change –it enables change

These changes are brought about by biocatalyst i.e. enzymes involved in the wear and tear and in regulation of chemical reactions that occur in plants and animals. Most of the metabolic reactions would occur at a rate very slow to support life, if enzymes were absent in the life system. Enzymes accelerate reactions by at least a million times. In addition to speeding up the reactions, they carry out the reactions in a much coordinated way thus playing a central role not only to individual reactions within a cell but also to the life of a cell as a whole.

In the mid 1700's, French scientist Rene-Antoine Reaumur and Italian biologist Lazzaro Spallanzani proved that gastric juices carry out the process of digestion. Their work laid the foundations which proved that vital reaction occurred outside the living organism. In the 1800's John R. Young added to the increasing knowledge about digestion by stating that gastric juice contains a strong acid, which was responsible for its digestive action. At the same time German physiologist Theodor Schwann discovered that gastric juice also contained a non-acid digestive substance called pepsin which was later shown to be an enzyme

Alcoholic fermentations were believed to be spontaneous reactions. In 1857, Louis Pasteur found that fermentation was caused by yeast cells digesting sugar for their own nourishment. In 1878 German physiologist Wilhelm Kuhn (1837-1900) coined the term "enzyme," meaning "in leaven," to describe the process of fermentation.

1n 1897, German scientist, Eduard Buchner, showed that intact living cells were not essential for carrying out any metabolic processes but thought there must be some small entities (called as enzymes) capable of converting sugar to alcohol. For this discovery Buchner won the 1907 Nobel Prize in Chemistry. In 1926, American biochemist James Sumner isolated the enzyme urease from the jackbean. The enzymes pepsin and trypsin were isolated by the American biochemist John H. Northrop.

A wide variety of chemical based processes are used today in the industry and a majority of them have drawback from economic and environmental point of view. Chemically synthesized drugs and other healthcare products have certain side effects which show their effects in the long run. Virtually all drawbacks can be eliminated using enzymes. With advent of incorporation of metabolic capacity of living system in many chemical industries, role of isolated enzymes or even living cells as a whole have found profound application. The advent of biotechnology has fueled a quantum leap in industries today as it supports not only efficient but also speedy reactions. Speed throughout the whole innovation chain is a prerequisite for the success. The use of biotechnology today is heartening in the entire possible arena to meet the needs of mankind. The role of microbes and of course microbiologists and biochemists cannot be emphasized enough in promoting biocatalytic reactions for production.

Biocatalytic process differ from conventional chemical processes, mainly due to enzyme kinetics, protein stability under conditions and catalyst features derived from their role in the cells physiology, such as growth, enzyme activity or use of metabolic pathways for various reactions. Organic reactions catalyzed by microorganisms are referred to as microbial transformations, biotransformation or bioconversions. Enzymes produced by microbial cells and by all living organisms catalyze biotransformation reactions. In their natural functions, enzymes catalyze, and indeed control, anabolic and catabolic reactions necessary to life processes. This chapter deals with bioconversion and the role of enzymes in the process.

4.2 BIOCONVERSION AND BIOTRANSFORMATION

A biotransformation can be defined as the specific modification of a defined compound to a defined product with structural similarity, by the use of biological catalysts. Bains (1993) describes the biocatalyst as an enzyme, or a whole, dead microorganism (that contains an enzyme or several enzymes). There is a very subtle difference between biotransformation and bioconversion. Bioconversion utilizes the catalytic activity of living organisms and hence can involve several chemical reaction steps. A living organism will be continuously producing enzymes and hence bioconversions often involve enzymes that are quite unstable. The properties of biotransformation and bioconversions are very similar and in many cases are cited as interchangeable (Walker and Cox 1995). Commercially the most utilized production process is fermentation. However, fermentation, where the product of metabolic activity often bears no structural resemblance to the pool of compounds given to the microorganisms as nutrients; is significantly different from biotransformation and bioconversions.

The terms bioconversion and biotransformation have been used as synonyms. Biotransformation is the chemical modification (or modifications) made by an organism on a chemical compound. If the modification results in a formation of mineral compounds like CO_2, NH_3^+ or H_2O, the biotransformation is called mineralization. Biotransformation means chemical alteration of substances such as nutrients, amino acids, toxins, or drugs in the body.

Various sources have given various definitions of biotransformation, as seen below:

Prescott *et al*, 2002, says that biotransformation or microbial transformation is "The use of living organisms to modify substances that are not normally used for growth."

McGraw-Hill Companies, 2003, stated that Biotransformation is "The series of chemical reactions that occur in a compound, especially a drug, as a result of enzymatic or metabolic activities by a living organism"

Stephenson et al, 2006, stated that biotransformation is the "Chemical conversion of a substance that is mediated by living organisms or enzyme preparations derived there from".

National Library of Medicine and Monosson, 2007, says that "Biotransformation is the process whereby a substance is changed from one chemical to another (transformed) by a chemical reaction within the body."

The American Heritage Dictionary of the English Language has given the biotransformation as a "Chemical alteration of a substance within the body, as by the action of enzymes."

In biology, Bioconversion has two meanings. The first one is known as biotransformation and uses microorganisms to carry out a chemical reaction which is costly or not feasible by any nonbiologically means. The microorganisms convert a substance to a chemically modified form *viz*. the bioconversion of Progesterone to 11-alpha-Hydroxyprogesterone by *Rhizopus nigricans*.

The second is the conversion of organic materials, such as plant or animal waste, into products or energy sources by biological processes or agents, such as certain microorganisms or enzymes. New cellulosic ethanol conversion processes have enabled the variety and volume of feedstock that can be biologically converted to expand rapidly. Materials derived from plant or animal waste such as paper, auto-fluff, tires, fabric, construction materials, municipal solid waste (MSW), sludge, sewage, etc can be used as raw material for conversion.

Having colonized virtually every environment, bacteria have evolved enzymatic solutions for a wide range of metabolic biochemical transformations. Such specific enzymes have found their way in solving many industrially viable reactions for production of desired compounds From a practical perspective, use of microbial enzymes have found wide applications. Microorganisms being critical to nutrient recycling in ecosystems act as decomposers, can fix elements like nitrogen and can also invade and disturb the metabolism of host organisms to the extent of being fatal. Use of microorganisms in brewing, baking, food-making processes and industrial bioconversion processes is a manifestation of their enzymatic activity.

Biotransformation of various pollutants is a sustainable and cheaper way to clean up contaminated environments. These bioremediation and biotransformation methods use the naturally present, microbial catabolic activity to degrade, transform or accumulate a variety of compounds *viz*. hydrocarbons polychlorinated biphenyls (PCBs), polyaromatic hydrocarbons (PAHs), oil, radionuclides, pharmaceutical substances, etc.. Major advances in the techniques have enabled analyses of environmentally relevant microorganisms providing helpful insights into biotransformation and biodegradative pathways and the ability of organisms to adapt to ever changing environment.

Many microbes have capacity to catabolically degrade compounds. Because of this property microorganisms naturally remove the contaminants and pollutants from the environment.. New methodological breakthroughs of Environmental Microbiology, genome-based global studies provide unprecedented in silico views of metabolic and regulatory networks, as well as new ideas to the evolution of biochemical pathways relevant to biotransformation and to the molecular adaptation strategies to changing environmental conditions. These new approaches have increased our understanding of the importance of different pathways and regulatory networks to carbon flux in particular environments and for certain compounds and they are speeding the development of bioremediation technologies and biotransformation processes. Also there is other approach of biotransformation called enzymatic biotransformation.

Novel catalysts can be procured from metagenomic libraries and DNA sequence based libraries. Our increasing capabilities in adapting the catalysts to specific reactions and process requirements by rational and random mutagenesis broaden the scope for application in the fine chemical industry, but also in the field of biodegradation. In many cases, these catalysts need to be exploited in whole cell bioconversions or in fermentations. This will help in understanding strain physiology and metabolism. Rational approaches to the engineering of whole cells will be a boost in the areas of biotechnology and synthetic biology.

4.2.1 ADVANTAGES OF BIOTRANSFORMATION:

The advantages of using biocatalysts are their ability to operate at near neutral pH, ambient temperatures and atmospheric pressures. Industrially useful chemistry often requires extremes of such conditions. Biocatalysts display very high reaction specificity, enantiomer- specificity and regio-specificity. This enantiomer of biotransformation has found many desirable application e.g. thalidomide that was produced as a racemate; had only one enantiomer that was effective safe anti-nausea agent. The other enantiomer produces teratogenic side effects in pregnant women. This concept of developing specific single enantiomer that not only protects against any adverse side effects of an undesired enantiomer but concentrates the activity of compound with only one active enantiomer has brought a revolution in the industrial world, although the US FDA policy allows pharmaceutical companies to market chiral compounds as racemic mixtures or single enantiomer.

The advent of movement not to use animal sources in production of natural sources was partly due to the perception amongst consumers of the transmission of animal diseases such as the encephalopathies (Bovine spongiform encephalopathy (BSE), scrapies, Kuru and Creutzfeld-Jacob Syndrome). Growth hormone purified from cadaver pituitaries was used to treat dwarfism but has been identified as the cause for the high occurrence of Creutzfeld-Jacob syndrome in patients (growth hormone is now produced by a recombinant *Escherichia coli)*; (Walker and Cox 1995). Many such survey and research work has high-lightened the need to move away from animal derived products. The production of compounds by plant and microbial source is therefore of great interest to the biological industries. There is therefore a huge potential for biotransformation processes, using microbial and plant enzymes to produce "nature-identical" compounds, replacing traditional animal based products.

Table 4.1 - Advantages and disadvantages of using whole cells and isolated enzymes for reactions

	Whole cells	Isolated Enzymes
Advantages	1) Cheap 2) Cofactors present.	1) Simpler equipments 2) Simpler work-up 3) Less contamination from other enzymes 4) Higher tolerance to organic solvents.
Disadvantages	1) High dilution, 2) Complex work-up, 3) Side–reactions caused by other enzymes present.	1) Expensive 2) Addition of cofactors necessary where required.

4.3 ENZYMES

Enzymes are mainly proteins, which catalyze, that means increase the rates of any chemical reactions. The concept of biocatalyst is very wide and includes enzymes, crude cell extracts, viable plant, animal and microbial cells and intact non-viable microbial cells.

In enzymatic reactions, the substrates are converted into different molecules, called the products. Almost all biological processes in a cell need enzymes to occur at different rates. The set of enzymes in a cell determines which metabolic pathways occur in that particular cell. Their importance to life is understood by the fact that many severe or fatal genetic diseases involve a missing or defective enzyme. One could design a drug which attaches to or occupies the active site of a target organism's enzyme. For example, penicillin destroys the enzyme required for the synthesis of peptidoglycan which is present in the bacterial cell walls. HIV protease inhibitors target the viral enzyme protease in AIDS treatment.

Enzymes like catalysts work by lowering the activation energy $(E_a^{\ddagger})$ for a reaction, thus dramatically increasing the rate of the reaction. Enzymes are not utilized by the reactions they catalyze, nor do they change the equilibrium of these reactions. However, enzymes do differ from most other catalysts as they are more specific. Enzymes are known to catalyze about 4,000 biochemical reactions.

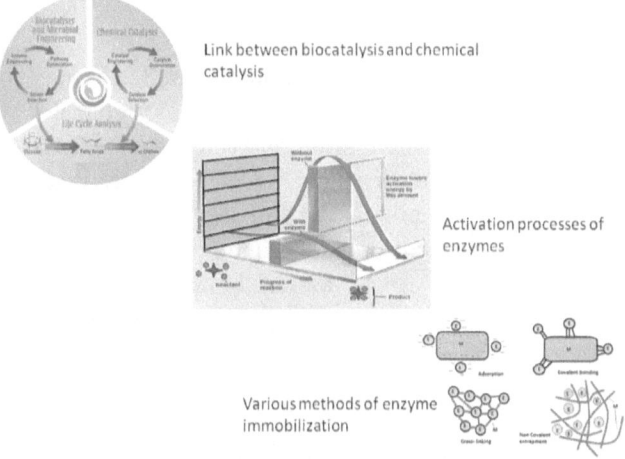

Link between biocatalysis and chemical catalysis

Activation processes of enzymes

Various methods of enzyme immobilization

Fig 4.1:- Enzyme mechanism and its role

As enzymes are proteins, they also get affected by the properties of proteins. Their activity and/or structure are/will be affected by changes in conditions such as temperature and pH. The rate of an enzyme-catalyzed reaction is also affected by the concentration of the enzyme and the concentration of the substrate. Other than substrate concentration, pH and temperature, there are other molecules which may affect enzyme activity such as inhibitors and activators. Inhibitors are molecules that decrease enzyme activity while activators are molecules that increase activity. Many drugs and poisons can act like enzyme inhibitors. Some enzymes are used for commercial purposes such as in the synthesis of antibiotics, whereas, some household products also use enzymes to speed up biochemical reactions.

4.3.1. ENZYME STRUCTURE

Enzymes are polymers of amino acids. They range in size from 1×10^4 Daltons to 1×10^6 Daltons with majority being in the 10^5 range. Some enzymes have extra co-molecules associated with them, which help in the reaction they carry out. The protein portion of an enzyme is called the apoenzyme. A cofactor is the non-protein part of an enzyme. Cofactors can be loosely bound, coenzymes or tightly bound, prosthetic groups. The complete enzyme (apoprotein + cofactor) is termed the holoenzyme.

4.3.2. CLASSIFICATION AND GENERAL PROPERTIES OF ENZYMES

The word "enzyme" was derived from Greek "in yeast" and was first introduced by Kuhn as a term for the substances formed in plants or animals which had previously been called "soluble" or "unorganized" ferments. As many activities previously regarded as characteristic of certain organisms have been found to be due to the activity of these enzymes, the conception of enzyme action has broadened. Until now the term enzyme was applied to all organic catalysts formed in plant or animal cells. Bacterial cells contain soluble enzymes within the cytosol and particulate enzymes bound to membrane structures. These are extracellular enzymes and intracellular enzymes. Many microbial enzymes are constitutive in nature i.e. they are always produced by the growing cell. If the enzymes are not constitutive, their formation in microorganisms can sometimes be induced by the substrate of interest, or by related compounds. Although each enzyme is generally supposed to be a definite chemical substance, the identification and classification of enzymes are based upon the changes which they bring about.

Enzymes are generally named by adding a suffix "-ase" to the primary name of the substrate molecule it is acting upon. For instance, Lipase catalyzes the hydrolysis of a lipid triglyceride. Sucrase catalyzes the hydrolysis of sucrose into glucose and fructose. There were a few enzymes discovered before this naming system was devised. These enzymes are known by common names such as pepsin, trypsin, and chymotrypsin which catalyze the hydrolysis of proteins.

The latest systematic nomenclature system known as the International Enzyme Commission (IEC) system is based upon the type of reaction catalyzed. There are six major groups of enzymes in this system as shown in table below.

Table 4.2 Different groups of enzymes and their activities

Group Name	Type of Reaction Catalyzed
Oxidases or Dehydrogenases	Oxidation- reduction reactions
Transferases	Transfer of functional groups
Hydrolases	Hydrolysis reactions
Lyases	Addition to double bonds or its reverse
Isomerases	Isomerization reactions
Ligases or Synthetases	Formation of bonds with ATP cleavage

4.3.3. ENZYME FUNCTIONS

Enzymes are "biological catalysts. Enzymes are very specific in nature. Each enzyme can act to catalyze only selective chemical reactions and only with selective substrates. An enzyme- substrate reaction has a lock-key reaction. An enzyme, being the key, has a certain structure or multi-dimensional shape that matches a specific section of the "substrate". Once these two components come together, certain chemical bonds within the substrate molecule change much as a lock is released and the enzyme is free to execute its duty once again. Enzymes often increase the rate of a chemical reaction between 10 and 20 million times.

Enzymes as proteins and biocatalyst has an important role in biotransformation

Enzymes, like other proteins, consist of long chains of amino acids held together by peptide bonds. They are present in all living cells, where they perform a vital function by controlling the metabolic processes. Enzymes take part in the breakdown of food materials into simpler compounds. They are found in the digestive tract eg. pepsin, trypsin and peptidases breakdown proteins into amino acids, lipases split fats into glycerol and fatty acid, and amylases breakdown starch into simple sugars.

Enzymes are biocatalyst and without being consumed in the process, they can speed up chemical processes. After the reaction is complete, the enzyme is released again, ready to start another cycle in the reaction. In principle most catalysts have a limited stability, and over a period of time they lose their activity and are not usable again. Based on above mentioned understandings and knowledge about enzymes produced by living systems, there has been efforts to create synthetic molecules called artificial enzymes that also also display enzyme-like catalysis.

4.3.4. BIOLOGICAL FUNCTION:

Enzymes perform a wide variety of functions inside living organisms like signal transduction, cell regulation, generate movement etc. Enzymes are also involved in more exotic functions, such as luciferase enzyme which helps in generating light in fireflies. Viruses also contain enzymes for infecting cells, such as the HIV integrase and reverse transcriptase.

An important function of enzymes is in the digestive systems of animals. Enzymes like amylases and proteases break down large molecules into smaller ones. Different enzymes digest different food substances. In ruminants which have herbivorous diets, microorganisms present in the gut produce another enzyme called cellulose which helps in breaking down the cellulose from cell walls of plant fibres. Several enzymes can work together in a specific order, creating metabolic pathways. The network of metabolic pathways within each cell depends on the set of functional enzymes that are present.

4.3.5. INDUSTRIAL APPLICATIONS

Many industrial enzymes originate from microorganisms in the soil. One micro-organism contains over 1,000 different enzymes. A long period of trial and error in the laboratory is required to isolate the microorganism for producing a specific type of enzyme. When the desired microorganism has been isolated, it can be mutated or transformed so that it is capable of producing the desired enzyme at higher yields. With the latest technological advancements of, it is possible to produce enzymes economically and in unlimited quantities. The end product of fermentation is a broth from which the enzymes are procured. This broth is then either centrifuged or filtered to remove the solid particles, that contains the residues of microorganism and raw materials which can be used as natural fertilizer. The enzymes are then used for various industrial applications.

Enzymes are used largely in the chemical industry and other industrial applications when extremely specific catalysts are required. Protein engineering is one such area of research and involves attempts to create new enzymes with novel properties, either through rational design or in vitro evolution. These efforts have begun to succeed, and a few enzymes have now been designed to catalyze reactions that do not normally occur in nature.

Microbial enzymes have been found to exhibit two advantages over the animal and plant enzymes. Firstly, they are economical and can be produced on large scale within the limited space and time and can be easily extracted and purified. Secondly, there is an advantage in using enzymes produced by microorganisms.

1. They are capable of producing a wide variety of enzymes,
2. They can grow in a wide range of environmental conditions,
3. They show genetic flexibility that is why they can be genetically manipulated to increase the yield of enzymes, and
4. Their generation times are short.

More than 2,000 enzymes have been isolated and characterized, out of which about 1,000 enzymes are used for various applications. Among them 50 microbial enzymes are used in various industrial applications.

4.3.6. IMMOBILIZED ENZYMES

Enzymes are widely and largely used commercially in various industries like food and brewing. Enzymes like Protease are used in 'biological' washing powders to increase the breakdown of proteins in stains like blood and egg. Enzymes such as Pectinase are used to produce and clarify fruit juices. With the use of various enzymes, various problems arise such as; enzymes being water soluble, they are hard to recover. Some enzymes are costly and impossible to obtain on a large scale. In some cases, the product may inhibit the activity of the enzyme by a process known as feedback inhibition.

To prevent these problems, enzymes can be immobilized by fixing them to a solid surface. This has a number of commercial advantages:

- The enzyme is easily removed.
- The enzyme can be packed into columns and used over a long period.
- Speedy separation of products reduces feedback inhibition.
- Thermal stability is increased allowing higher temperatures to be used.
- Higher operating temperatures increase rate of reaction.

There are four principal methods of immobilization currently in use:

- Covalent bonding to a solid support
- Adsorption onto an insoluble substance
- Entrapment within a gel
- Encapsulation behind a selectively permeable membrane

Finally it can be said that for synthesis of Mandelic, use of enzyme offers great potential. The enzyme Nitrilase catalyzes the hydrolysis of nitriles to carboxylic acids and ammonia. This in process there is no formation of free amide intermediates. They work at neutral pH and ambient temperature without affecting other labile functional groups such as esters and amides. The Nitrilase enzyme has a very high substrate specificity toward aromatic nitriles (benzonitrile, 3-cyanopyridine and 4-cyanopyridine) and unsaturated aliphatic nitrile (acrylonitrile). It is also able to catalyze saturated aliphatic nitriles (like acetonitrile, propionitrile, butyronitrile and isobutyronitrile) nitrile and arylacetonitrile (phenylacetonitrile and indole-3-acetonitrile). They are used for the manufacture of Mandelic acid, nicotinic acid and the detoxification of cyanide waste. With the help of Nitrilase obtained from microbial and plant sources, the conversion will be much cheaper compared to extraction from almonds and chemical conversion methods which may result in toxic substances to be released out.

4.4 CONCLUSION

With the increased and successful use of enzymes in most of the industrial sector, bioconversion reactions are used to obtain products at a cheaper cost and a faster rate. These enzymes being eco friendly tend to minimize the release of toxic and harmful products. The conversion of mandelonitrile to Mandelic acid can be carried out using such enzymes which will prevent the release of cyanide waste during the process.

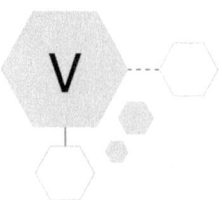

NITRILASES

Content:

5.1 INTRODUCTION TO NITRILASE

There are naturally occurring enzymes which have great potential for use in industrial chemical processes for the conversion of nitriles to a wide range of useful products and

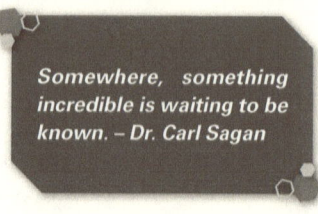

Somewhere, something incredible is waiting to be known. – Dr. Carl Sagan

intermediates. Such enzymes include Nitrilase which are capable of converting nitriles directly to carboxylic acids. Nitrilases have commercial utility as biocatalysts for use in the synthesis of enantio-selective aromatic and aliphatic amino acids or hydroxy acids.

Nitrilases enzymes are found in a wide range of mesophilic micro-organisms, including species of *Bacillus, Nocardia, Bacteridium, Rhodococcus, Micrococcus, Brevibacterium, Alcaligenes, Acinetobacter, Corynebacterium,, Fusarium and Klebsiella*. Additionally, there are thermophilic Nitrilases which exist in bacteria.

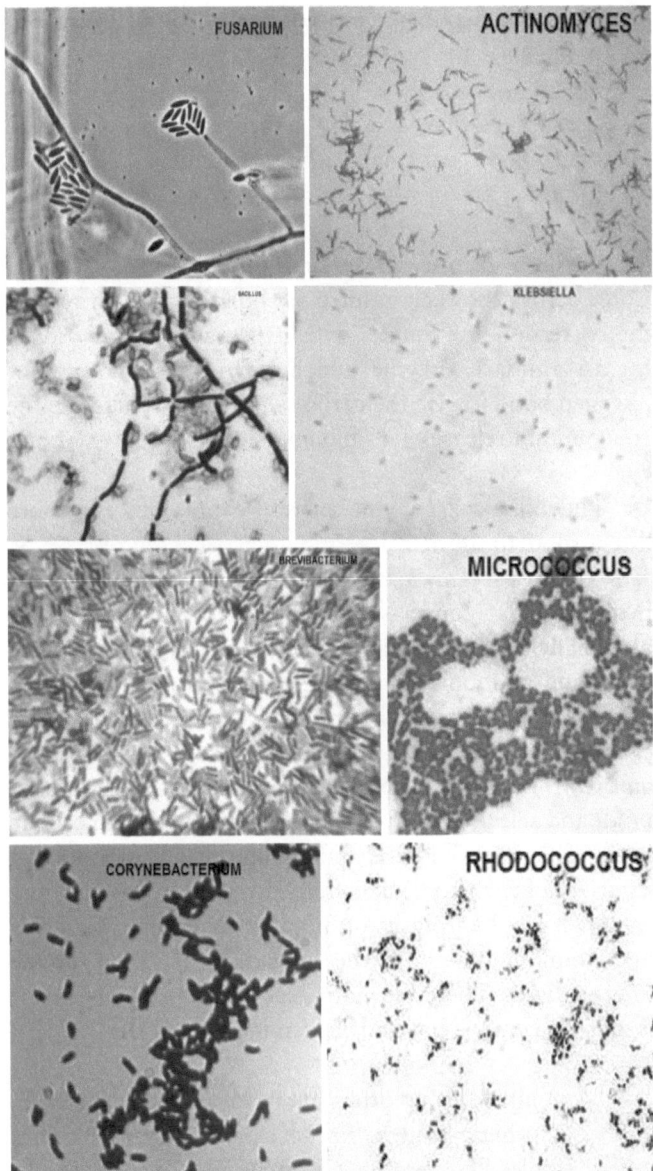

Figure 5.1: Various microorganisms rich in Nitrilase
(Courtesy: http://www.Pseudomonas.com)

There are two major ways of conversion of a nitrile to an analogous acid

(1) Nitrilase catalyzed direct hydrolysis of a nitrile to a carboxylic acid with the concomitant release of ammonia; or

(2) Nitrile hydratase catalyzed addition of a molecule of water across the carbon-nitrogen bonding system to give the corresponding amide, which then acts as a substrate for an amidase enzyme which hydrolyzes the carbon-nitrogen bond to give the carboxylic acid product with the concomitant release of ammonia.

The Nitrilase enzyme therefore provides the more direct route to the acid.

Nitrilases (EC 3.5.5.1) are an important class of hydrolase that converts naturally occurring, as well as artificially derived, nitriles to the corresponding carboxylic acids and ammonia. The transformation acts through a cysteine residue in the active site and requires no metal cations or other cofactors. Because of their inherent enantio- and regio-selectivities, Nitrilases are attractive as 'green', mild, and selective catalysts for setting stereogenic centers in fine chemical synthesis and enantiospecific synthesis of a variety of carboxylic acid derivatives. Nitrilase enzymes catalyze the hydrolysis of nitriles to carboxylic acids and ammonia, without the formation of "free" amide intermediates. There are many reasons as to why Nitrilase is gaining commercial and academic importance:

1. A nitrile group offers many advantages in devising synthetic routes for developing new medicinal molecules, because it is often easily introduced into a molecular structure and can be carried through many processes as a masked acid or amide group.

Hence, this enzyme is used in the manufacture of fine chemicals. This is important because the chemicals show enantioselectivity.

2. The structure of Nitrilase having a glu, cys, lys catalytic triad offers versatile chemistry.

3. The chemo selective biocatalytic hydrolysis of nitriles represents a valuable alternative because it occurs at ambient temperature and near physiological pH.

4. The proper role of Nitrilase in metabolism is not well characterized.

5. Microbes utilize Nitrilases for their survival in highly toxic environment. Nitrilases are highly useful for environmental detoxification. Cyanide represents a widely applicable C1-synthon (cyanide is one of the few water-stable carbanions) which can be employed for the synthesis of a carbon framework. However, further transformations of the nitrile thus obtained are impeded due to the harsh reaction conditions required for its hydrolysis using normal chemical synthesis procedures. The use of enzymes to catalyze the reactions of nitriles is attractive because Nitrilase enzymes are able to effect reactions with fewer environmentally hazardous reagents and by-products than in many chemical methods.

Nitrilases were the first nitrile-metabolizing enzyme to be discovered (Thimann and Mahadevan 1964) and it is known to convert indole- 3-Acetonitrile to indole-3-acetic acid (an auxin) in plants. The discovery of microbial Nitrilase was carried on much later.

Nitrilase activity in metabolism occurs in plants, animals, fungi and certain prokaryotes. Apart from synthesizing natural products, this enzyme Nitrilase are also involved in post translational modifications.

Nitriles may also be converted to acids via bioenzymatic pathways involving nitrile hydratases (EC 4.2.1.84) and amidases (EC 3.5.1.4), the former catalyzing the hydration of nitriles to the corresponding amides, followed by their conversion to acids by the later. The Nitrilases can also act as nitrile hydratases and vice versa, and that Nitrilases of extremophiles are known with interesting stabilities. Unfortunately, <20 Nitrilases have been reported in the scientific and patent literature because of stability or specificity.

The nitriles occur widely in the environment. In nature, they are produced by many plants in various forms such as Cyanoglycosides, Cyanolipids, Ricinine, Phenylacetonitrile, etc. Although most of the nitriles are toxic, mutagenic, and carcinogenic in nature, they are also important intermediates in the organic synthesis of amines, amides, amidines, carboxylic acids, esters, carbonyl compounds, and heterocyclic compounds. Hence, Nitrilases, along with other nitrile-converting enzymes, are becoming important biocatalysts with potential applications in different fields of organic reactions.

5.2 STRUCTURE OF NITRILASE

Nitrilases belong to a subfamily of the carbon-nitrogen hydrolase super-family. Members of the super-family share an α-β-β-α sandwich fold, similar topologies, and characteristic sequence motifs. All known Nitrilase homologues are at least di-mers. There are four crystal structures of homologues of which two are dimers and two are tetramers with 222 symmetry. Only one has known enzyme activity – a carbamylase.

Several published work has shown three forms built around a helical symmetry – two are terminating spirals with 14 and 18 subunits respectively. The 18-mer changes to an open helix at pH5.4. Several studies have shown that in case of *B. pumilus*, an activity increase accompanies the pH dependent increase in size and that "destructive" mutation of residues in the interfaces which associate to form helices abolishes activity.

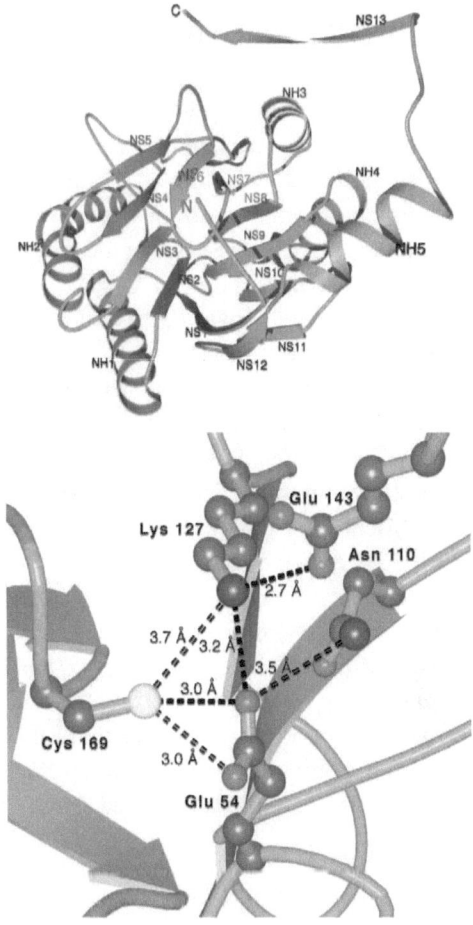

Figure 5.2: Structure of Active site of Nitrilase enzyme along with its ribbon drawing

Most determinations of oligomeric state are based on gel-filtration and are unreliable – but the numbers range from 2-18 subunits for all known Nitrilases. Nagasawa et al (1990) has shown that di-mers are inactive until they oligomerize (in *Rhodococci*).

5.3 PHYSICO-CHEMICAL PROPERTIES OF NITRILASE

The full length amino acid sequences are from about 333 amino acids to about 366 amino acids, It has sggregation and activity as homo –multimers of about 2 subunits to about 16 subunits, There is presence of a catalytic triad of the consecutive amino acids Glu-Lys-Cys,

Its pH optima range from pH 5 to about pH 9, and it has an optimum pH near 7.0

Temperature optima is from about 0^{0}C to about 100°C, being most optimal between 40^{0}C to 50^{0}C. However, it slowly becomes inactivated at $35°$ C and above in air.

Nitrilase shows energy of activation of 14,000 Cal per mole, and a Michaelis constant for the substrate Indole Acetonitrile (IAN) of 5.1×10^{-5} at $25°$ C.

It is susceptible to SH-reagents. The conversion of Indole Acetonitrile (IAN) to Indole Acetic Acid (IAA) and ammonia is quantitative, requires no oxygen, and does not yield free amide as an intermediate product.

Starch gel electrophoresis indicates that all activity is associated with a single protein fraction, so that the two stages of hydrolysis are probably carried out by one and the same enzyme.

5.4 THE NITRILASE SUPER-FAMILY

Members of the Nitrilase super-family perform an array of carbon-nitrogen bond hydrolysis and condensation reactions in natural products biosynthesis and protein modification. The Nitrilases super-family as consisting of 12 families of amidases, N-acyltransferases and presumptive amidases, in addition to the family of plant and bacterial Nitrilases for which the super-family was named. A novel Glu-Lys-Cys catalytic triad, found at the crystallographically defined Nit active site of enzyme, was postulated to constitute the catalytic residues for all members of the super-family. Recent experimental results confirm the essentially of the catalytic triad residues and specify the biochemical functions of additional branches and sub-branches of the Nitrilase super-family.

Nitrilases are classified based on their substrate specificity as aliphatic or aromatic Nitrilases. Depending on the activity and source, 13 branches of enzyme Nitrilases have been named as follows:

Branch 1: Nitrilase

Members of Nitrilase branch (EC 3.5.5.1) are found in plants, animals. In some fungi e.g. *C. elegans*, *Saccharomyces cerevisiae's* and many types of bacteria; *NIT1* gene responsible for producing this enzyme is inactivated.

The best evidence that Nitrilase functions *in vivo* to convert Indole Acetonitrile to the plant growth factor Indole-3-Acetic Acid (auxin). Nitrilases from bacterial are used for biochemical syntheses as well as for environmental remediation. Whether bacterial Nitrilases primarily function in ecological relationships with plants or whether they benefit isolated microbes is not yet clear. Nitrilase I tolerates aromatic ring substituents in the ortho-, meta-, and para-positions of Mandelonitrile derivatives, and products are produced with high enantio-selectivities.

Branch 2: Aliphatic Amidase

Aliphatic amidases (EC 3.5.1.4) hydrolyze substrates such as the carboxamide side chains of glutamine and asparagine utilizing the cysteine conserved within the Nitrilase super-family.

Branch 3: Amino-terminal Amidase

This protein deaminates amino-terminal asparagine and glutamine residues to aspartate and glutamate. The N-end rule is a means by which the rate of ubiquitin-dependent protein degradation is regulated. The *S. cerevisiae* Nta1 protein deaminates amino-terminal asparagine and glutamine residues, which lead to more rapid rates of protein turnover.

Branch 4: Biotinidase

Biotinidases (EC 3.5.1.12) utilize specific amidase/esterase activity to release biotin from biotin-amide, biotin-lysin, biotin-peptide conjugates and biotin methyl esters. Biotinidase enzymes are the only amidases in the Nitrilase super-family known to prefer secondary amine substrates of the form R-C=O (NHR') as opposed to simple acid amides.

Branch 5: Ureidopropionase

The ureidopropionases (EC 3.5.1.6) are enzymes involved in the catabolism of pyrimidine bases and the production of alanine. Substrates of this enzyme are of the carbamylase type and the amine product is usually a non-standard amino acid.

Branch 6: Carbamylase

A variety of bacteria express hydrolases specificity for the decarbamylation of D-amino acids. These enzymes have been exploited in the production of semi synthetic ß-lactam antibiotics.

Branches 7 and 8: Glutamine-dependent NAD synthetase

Glutamine- dependent NAD synthetase (EC 6.3.5.1) is able to utilize glutamine as an ammonia source. Substrate specificity of Nitrilase-related proteins as glutamine amidases is not surprising given the specificity of the branch 2 and 3 enzymes. It remains to be seen how glutamine-dependent NAD synthetase may channel ammonia from the Nitrilase-related active site to the NAD active site.

Branch 9: Apolipoprotein N-acyltransferase

Branch 9 enzymes condense a fatty acid to the amino terminus of the modified cysteine residue.

Branch 10: Nit:

The branch 10 enzymes are implicated in the function of Fhit by virtue of Rosetta stone relationships. Nit was originally identified as an approximately 300 amino acid amino-terminal extension on fly and worm homologs of the human and murine Fhit tumor suppressor protein. . The Fhit active site of NitFhit has been characterized and the structure of worm NitFhit has been elucidated [12], but the Nit substrate, cell biology and relationship to tumor suppression are not known.

Branches 11:

Branches 11 and 12 contain distinct similarity groups with no characterized member. Branch 11 was classified as containing 13 sequences of unknown specificity. It is reported that the product of the *Pseudomonas AguB* gene is an *N*-carbamyl putrescine amidohydrolase that is related to β-ureidopropionases. This enzyme is thought to function in the metabolism of arginine into spermidine and succinate and it is reported that AguB is a new member of branch 11. These bacterial and plant enzymes, which now number more than 20, probably function as carbamylases for the production of putrescine or related amines.

Branches 12:

The branch 12 enzymes are implicated in protein post-translational modification by virtue of our observation that they are sometimes fused with the RimI N-terminal acetyltransferases. Branch 12 may contain Rosetta Stone proteins in that a distinctive nitrilase-related domain is found fused to an amino-terminal domain of approximately 210 amino acids.

Branches 13:

Branch 13 remains a bin of nonfused outliers, without a single clear prototype, that is expected to be classified into multiple branches once functional information is available for some of its sequences.

5.5 CYANIDE DEGRADING NITRILASES

Recombinant forms of three cyanide degrading Nitrilases (i) CynD from *Bacillus pumilus* C1, (ii)

CynD from *Pseudomonas stutzeri* and (iii) CHT from *Gloeocerospora sorghi*, were prepared after their genes were cloned with C-terminal hexahistidine purification tags and expressed in *Escherichia coli,* and the enzymes was purified using nickel-chelate affinity chromatography. The enzymes were compared with respect to their pH stability, thermo stability, metal tolerance and kinetic constants. The two bacterial genes, both cyanide dihydratases, were similar with respect to pH range, retaining greater than 50% activity between pH 5.2 and 8 and kinetic properties, having similar Km (6-7 Mm) and V(max) (0.1 mmol min (-1) mg (-1). They also exhibited similar metal tolerances. However, the fungal CHT enzyme had notably higher Km (90 Mm) and V (max) (4 mmol min (-1) mg (-1)) values. Its pH range was slightly more alkaline (retaining nearly full activity above 8.5), but exhibited a lower thermal tolerance. CHT was less sensitive to Hg $^{(2+)}$ and more sensitive to Pb$^{(2+)}$ than the CynD enzymes. These data describe, in part, the current limits that exist for using Nitrilases as agents in the bioremediation of cyanide-containing waste effluent, and may help serve to determine where and under what conditions these Nitrilases may be used.

5.6 NITRILE HYDRATASE AND AMIDASE

In microorganisms (bacteria and fungi), for the sequential metabolism of nitrile compounds, two hydrolytic enzymes namely Nitrile hydratase and Amidase are responsible. These enzymes are not to be confused with Nitrilases. They work in tandem to convert a nitrile to a corresponding carboxylic acid. These enzymes are capable of utilizing aliphatic nitriles as the sole source of nitrogen and carbon. Nitrile hydratases (EC 4.2.1.84) catalyze the hydration of a nitrile to the amide, which subsequently can be converted to the carboxylic acid and ammonia by a corresponding amidase (EC 3.5.1.4).

5.7 DETECTING THE PRESENCE OF NITRILASE

There are several different types of assays which can be performed to test for the presence of Nitrilase activity in a sample or to test whether a particular polypeptide exhibits Nitrilase activity. Assays can detect the presence or absence of products or by-products from chemical reaction catalyzed by a Nitrilase e.g., the presence of Nitrilase activity can be detected by the production of α-hydroxy acids or α-amino acids from, respectively, cyanohydrins or aminonitriles, and the level of Nitrilase activity can be quantified by measuring the relative quantities of the reaction products produced.

In one aspect, Nitrilases of the invention can stereoselectively hydrolyze nitriles or cyanohydrins into their corresponding acids and ammonia. The various methods for the detection of Nitrilase are:

a. Chemical methods include carrying out bioconversion using cell extract and adding 1ml of 0.011% sodium nitroprusside and l ml of 20mM sodium hypochlorite to the product formed. The solution is heated in a boiling water bath for 10 min and the optical density is measured at 640 nm. But there are limitations to this method as the buffer is not stable, thus it is not used for a large number of libraries and interference is caused by other proteins.

b. In another method $CoCl_2$ is used, which changes from pink to light yellow color after reacting with the bioconversion product (normally reacted with the ammonia released); which is determined spectrophotometrically at 375 nm. The drawback of this method is that it is not able to detect concentrations below 5mM.

c. Usually conventional methods such as HPLC, liquid chromatography-mass spectrometry, or gas chromatography are used. But these methods are tedious and time consuming.

d. Methods for efficient screening for Nitrilase producing organisms are required. These methods include the use of fluorescent probes. The lanthanide ions, such as Tb(III), Eu(III), Sm(III), and Dy(III), exhibit typical fluorescence. Salicylic acid is used to attach these lanthanide ions and they serve as a photon antenna. Nitrilase activity was first measured using non-fluorogenic probes, and product concentration was monitored by HPLC. This is a simple, rapid and high throughput fluorescence Nitrilase assay method with high specificity. The accuracy is higher than those of the other existing assay methods.

5.8 OCCURRENCE OF NITRILASES

Nitrilases are ubiquities and are produced by various living systems viz. microbe, plants, fungi and even some animals.

5.8.1. IN MICROBES

Several fungi have this enzyme; but they do not excrete it into the culture medium. About 20 Nitrilases have been identified in bacteria, half of them having been identified in sequenced genomes and not experimentally tested for their activity. Though the members of the super family have been identified also in archaea, but no archaeal enzyme with nitrile-hydrolyzing activity has been reported. Many bacteria and archea, especially those with an ecological relationship to plants and animals, encode members of the Nitrilase super family and utilize the enzymes for chemically similar nitrile or amide hydrolysis reactions or for condensation of acyl chains to polypeptide amino termini.

Nitrilases are found in a wide range of mesophilic micro-organisms (e.g. *Bacillus*, *Norcardia*, *Bacteridium*, *Rhodococcus*, *Micrococcus*, *Brevibacterium*, *Alcaligenes*, *Acinetobacter*, *Coryebacterium*, *Fusarium* and *Klebsiella*) and also in thermophilic bacteria.

Nitrilases from cultured species of the genera *Pseudomonas*, *Alcaligenes*, *Rhodococcus*, and *Acinetobacter* have previously been used in stereo-selective syntheses of substituted (R)-Mandelic acids, carbohydrate acids, (S)-α-phenylglycine, (S)-naproxen, and (S)-ibuprofen.

An immobilized Nitrilase has been used in many cases. A method for preserving immobilized or unimmobilized microbial cells having Nitrilase activity has been developed. This method also stabilizes the Nitrilase activity.

This is carried out by preparing aqueous suspensions containing at least 100 mM bicarbonate, carbonate, or carbamate salts. This limits microbial contamination of the stored enzyme catalyst, as well as stabilizes the desired Nitrilase activity. This has been carried out in *Acidovorax facilis* 72-PF-15 (ATCC 55747), *Acidovorax facilis* 72-PF-17 (ATCC 55745), *Acidovorax facilis* 72W (ATCC 55746), and many transformed microbial cells having Nitrilase activity.

5.8.2. IN FUNGI

The Nitrilase super family consists of thiol enzymes involved in natural product biosynthesis and post translational modification in plants, animals, fungi and certain prokaryotes. Nitrilases in the form of aminonitriles and cyanohydrins are found in fungi. Some pathogenic fungi have cyanide hydratases that allow them to infect plants that synthesize large amounts of defensive alkenyl glucosinates, which break down into isothiocyanates and nitriles. These fungi do no excrete it into the culture medium.

Plants, animals and fungi perform a wide variety of nonpeptide carbon nitrogen hydrolysis reactions using members of the Nitrilase superfamily. These Nitrilase and amidase reactions, which produce auxin, biotin, β- alanine and other natural products, and which results in deamination of protein and amino acid substrates, all involve attack of cyano or carbonyl carbon by a conserved cysteine.

5.8.3. IN PLANTS

This enzyme is not common in the plant kingdom. A screening done by Thimann and Mahadevan (1963) has revealed that out of 29 plants (from 21 families) tested, 19 showed no activity and only members of the Gramineae (grasses), Cruciferae (cabbage group and radish), and Musaceae (banana family) were clearly active. The enzyme has been partially purified from barley leaves.

Naturally occurring Mandelic acid is found when amygdalin (a cyanogenetic glycoside found in many plants including bitter almond, apricot, and wild cherry) is spirit by hydrolysis with hydrochloric acid, while amygdalin is broken down into glucose, benzaldehyde, and prussic acid (hydrogen cyanide) in the presence of sulfuric acid.

Nitrilases of plants are divided into two groups, depending if they are homologs Nitrilase 4 (NIT4) or Nitrilase 1 (NIT1) from *Arabidopsis thaliana*.

N1T1 is restricted to Brassicaceae family. It shows broad substrate specificity. It is also involved in glucosinolate catabolism.

N1T4 is widely distributed in the plant kingdom. It has high substrate specificity for β-cyanoalanine. It is also involved in cyanide detoxification.

Plants may be challenged by the highly toxic compound cyanide (hydrocyanic acid) from two sources:

1. Biosynthesis of the plant hormone ethylene. Biosynthesis of this hormone is strongly induced e.g. by wounding, during senescence and fruit ripening.

2. Wounding-induced breakdown of cyanogenic glycosides by the sequential action of β-glucosidases and α-hydroxynitrile lyases. This process is restricted to plants which contain cyanogenic glycosides (approx. 2600 known species)

Radish   Cabbage   Barley

Figure showing Plants rich in nitrilases

Bitter Almonds   Apricot   Wild Cherries

Fig 5.3: Plants rich in nitrilases and amygdalin

Plants are able to detoxify this cyanide via a two-step pathway. In the first step, cyanide is coupled to the amino acid L-cysteine, thus forming the non-proteinogenic amino acid β-cyano-L-alanine. By this reaction the cyanide is detoxified but the formed β-cyanoalanine is more or less useless for the plant. In the second step, therefore, the β-cyanoalanine is converted into the proteinogenic amino acids L-asparagine and L-aspartic acid.

In the past it was seen that NIT4 of the model plant *Arabidopsis thaliana* and the NIT4 homologs of *Nicotiana tabacum* can use β-cyanoalanine as substrate and that these enzymes form two products, namely asparagine and aspartic acid, at the same time (ref 36). In addition the β-cyanoalaninehydrolyzing activity found in *Lupinusangustifolius* is also due to a NIT4 homolog. In contrast to the NIT4 enzymes from *Arabidopsis thaliana* and *Nicotiana tabacum*, the lupine NIT4 produces much more asparagine than aspartic acid.

The primary component of bitter almond oil extract is benzaldehyde (a precursor for Mandelic acid synthesis) and can be extracted from a number of other natural sources in which it occurs, such as apricot, cherry, and laurel leaves, peach seeds and, in a glycoside combined form (amygdalin), in certain nuts and kernels.

5.8.4. IN ANIMALS

Mandelonitrile are reported to be present in arthropods i.e. when *Apheloria* a millipede is about to be attacked, mandelonitrile decomposes enzymatically (Nitrilase) into benzaldehyde and a poison (hydrogen cyanide). Amygdalin is known as laetrile, which has been used as an anti-cancer drug. But it is known that it does not kill cancerous cells selectively.

5.9 NITRILASE CATALYZED REACTIONS IN LIVING SYSTEMS

The Nitrilase directed catalysis includes double hydration of the nitrile to produce the corresponding carboxy acid without the formation of intermediate amide. The enzyme attacks the nitrile substrate covalently producing ammonia with the first water molecule addition. The acid is then produced with the regeneration of the enzyme and the addition of the second water molecule. The chemo-selective bio-catalytic hydrolysis of nitriles represents a valuable alternative because it occurs at ambient temperature and near physiological pH without affecting other labile functional groups such as esters and amides. The enzyme Nitrilase hydrolyzes many kinds of nitrile compounds such as aliphatic, aromatic, and heterocyclic mono- or di-nitriles.

Members of the Nitrilase super-family perform an array of carbon-nitrogen bond hydrolysis and condensation reactions in biosynthesis of natural products and protein modification. As mentioned above the Nitrilase super-family consists of 12 families of amidases, N-acyltransferases and presumptive amidases. A novel Glu-Lys-Cys catalytic triad has been found at the crystallographically defined Nit active site of enzyme, which was postulated to constitute the catalytic residues for all members of the super-family. Recent experimental results confirm the essentiality of the catalytic triad residues and specify the biochemical functions of additional branches and sub-branches of the Nitrilase super-family.

It was seen that three different Nitrilases lost 50–100% of their activity upon exposure to oxygen for 40 h, whereas their activity was fully retained under an inert argon atmosphere. This effect is ascribed to a reaction of oxygen, presumably with the catalytic cysteine residue. The ratio of the Nitrilase and nitrile hydratase activities of the enzyme is profoundly influenced by the electronic and stearic properties of the reactant.

Bearing in mind that plant, fungi, bacteria, algae, insects and sponges synthesize many organic nitriles; it is not surprising, that there are several biochemical pathways for nitrile degradation. From a preparative standpoint, the most interesting among them involves nitrile hydrolysis. The latter may take place via two different pathways:

- Single-step hydrolysis catalyzed by a Nitrilase, yielding the corresponding carboxylic acid, where as
- A two-step process involves formation of the carboxamide which is catalyzed by a nitrile hydratase followed by formation of carboxylic acid catalyzed by an amidase. Both types of nitrile-converting enzymes act via distinctively different mechanisms.

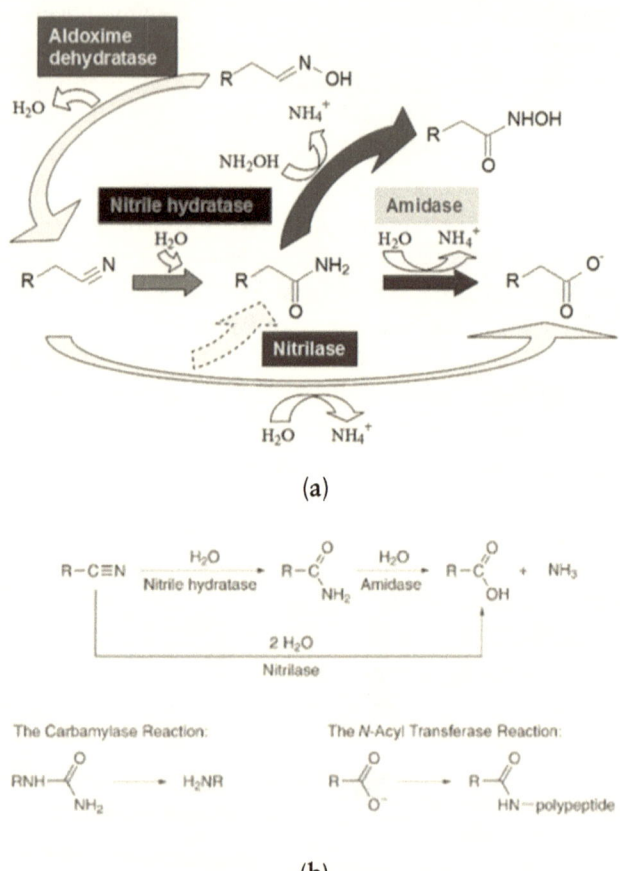

(a)

(b)

Figure - 5.4: Types of reactions carried out by
selected Nitrilase Super-family Members

Hence the Enzyme Pathways can be illustrated as

Nitrilase
R-CN + 2H→R-COOH + NH

Cyanide Hydratase
HCN + H-→HCONH→R-CN + 2H→R-COOH + NH

Nitrile Hydratase/Amidases

R-CN + -----R-CONH----------R-COOH + NH

5.10 NITRILASE CATALYZED CONVERSION OF MANDELONITRILE TO MANDELIC ACID

Nitrilases are commercially used for conversion of Mandelonitrile to Mandelic acid, nicotinic acid and the detoxification of cyanide waste. Apart from chemically procured Nitrilases; efforts have been on to use microbes and plants as source of Nitrilases i.e. because Nitrilase, which catalyses the hydrolysis of nitriles to the corresponding acids and ammonia, was originally discovered as a plant hormone involved in Indole acetic acid metabolic pathway.

5.11 PRESERVATION AND STABILIZATION OF IMMOBILIZED OR UNIMMOBILIZED MICROBIAL CELLS HAVING NITRILASE ACTIVITY

Microbial cells having Nitrilase activity can be stabilized by storing them in aquesous solution of inorganic slats. Aqueous solution of inorganic salts of carbamate, carbonate or bicarbonate prevents microbial contamination of the stored enzyme catalyst and also stabilizes the Nitrilase activity. These solutions and can also be used for storage of Nitrilase activity. Types of inorganic salts that can be used are Sodium, potassium, ammonium etc. Sometime the combination of salts can be used. Various methods can be used for immobilization of microbial cells having Nitrilase activity such as use of algenate gel, poly acrylamide gel, adsorption or attachment to ion exchange resins etc. Choice of inorganic salt used also depends on the method of immobilization used.

5.12 EVOLUTIONARY ANALYSIS OF NITRILASES

Nitrilases are important in the biosphere as participants in synthesis and degradation pathways for naturally occurring, as well as xenobiotically derived, nitriles. Because of their inherent enantioselectivity, Nitrilases are also attractive as mild, selective catalysts for setting chiral centers in fine chemical synthesis. Robertson *et al* (2004) characterized 137 unique Nitrilases, by screening more than 600 biotope-specific environmental DNA (eDNA) libraries. Using culture-independent means, phylogenetically diverse genomes were captured from entire biotopes, and their genes were expressed heterologously in a common cloning host. Nitrilase genes were targeted in a selection-based expression assay of clonal populations numbering 106 to 1010 members per eDNA library. A phylogenetic analysis of the novel sequences discovered revealed the presence of at least five major sequence clades within the Nitrilase subfamily. Using three nitrile substrates targeted for their potential in chiral pharmaceutical synthesis, the enzymes were characterized for substrate specificity and stereo specificity. A number of important correlations were found between sequence clades and the selective properties of these Nitrilases. These enzymes, discovered using a high-throughput, culture-independent method, provide a catalytic toolbox for enantiospecific synthesis of a variety of carboxylic acid derivatives, as well as an intriguing library for evolutionary and structural analyses

5.13 INDUSTRIAL APPLICATIONS OF NITRILASE

Nitrilases may be used in the synthesis of industrially important carboxylic acids which are otherwise produced by chemical methods requiring extreme conditions of temperature and pH. The Nitrilase of *R. rhodochrous* J1 has been used to hydrolyse 3-cyanopyridone to nicotinic acid, a vitamin used in animal feed and medicine.

Nitrilases may also be used in the bioremediation of land and water contaminated with toxic nitrile compounds. These compounds enter the environment from a variety of sources. In industry, acetonitrile is used as a solvent while acrylonitrile is used in the synthesis of plastics. Nitrile compounds such as bromoxynil are used as herbicides, and cyanide is used to manufacture plastics and the extraction of precious metals as well as being released to the environment as a toxic by-product of mining, metal finishing and organic chemical industries. The Nitrilase of *K. pneumoniae* sp. *Ozaenae* is seen to be highly specific for the herbicide bromoxynil. This enzyme has been expressed in plants, and confers herbicide resistance to transgenic lines. the nitrilase from the thermophilic bacterium *B. pallidus* strain Dac521 has a broad substrate range and is extremely stable at high temperatures, while the nitrilase of the hyperthermophilic archeon *Pyrococcus abyssi* shows catalytic activity at high temperatures and across a wide pH range.

5.14 CONCLUSION

Until very recently, the commercial availability of this valuable enzyme has been very limited. But now, screening sets of Nitrilases are available world wide, allowing the rapid identification of Nitrilase for any given application. The enzymes are manufactured for easy use, including all components necessary for activity. Continuous research is ongoing to further expand this platform. The range of possible Nitrilase applications has been recently widened but in many cases the parameters of the reactions need to be changed to establish viable industrial processes. To achieve this goal, several methods have been used, primarily in screening for enzymes from new sources, improving the enzyme, modification of the substrate structure and variation in various process parameters. New Nitrilases have been obtained by various advanced methods like genome mining and the metagenomic studies. Protein engineering has revealed many targets for altering substrate specificity, activity or enantioselectivity, and for changing the acid: amide ratio in the Nitrilase product. In addition, enzymes made for specific applications can be developed using directed evolution methods when rate and selectivity are not yet optimal. It is clear that

Nitrilases can be of great importance in organic chemistry and because of their increasing availability and especially their ease of use. Variation in the structure of the substrate proved to be another method to modify the enantio-selectivity and chemo-selectivity of the reactions. The use of Nitrilases will be increased further if they are applied in combination with other enzymes or chemical catalysts without intermediate isolation. Such processes have been developed for the synthesis of (S)-Mandelic acid, (S)-mandeloamide or glycolic acid.

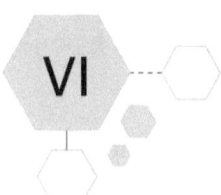

MANDELIC ACID: CHEMICAL ANALYSIS

Content:

6.1 INTRODUCTION

According to web definition:

"*An analytical technique is a method that is used to determine the concentration of a chemical compound or chemical element.*"

Research is to see what everyone else has seen, & think what nobody has thought

Common analytical techniques used for analysis includes gravimetric, Titrimetric etc whereas very advanced techniques encompasses use of highly specialized instrumentation.

The most common techniques used in analytical detection are as follows.

> Titrimetry, based on the quantity reagent of the needed to react with the analyte
> Electro-analytical technique, including potentiometry and voltammetry
> Spectroscopy based on the interaction of the analyte with electromagnetic radiation
> Chromatography in which the analyte is separated from the rest of the sample so that it may be measured without interference from other compounds
> Microscopy
> Bioanalysis
> Radioanalytical chemistry

Generally organic compounds consist of carbon atoms, hydrogen atoms and functional groups. The valence of carbon is 4, and hydrogen is 1, functional groups are generally 1. From the number of carbon atoms and hydrogen atoms in a molecule the degree of unsaturation can be obtained. Many but not all structures can be envisioned by the simple valence rule that there will be one bond for each valence number. The knowledge of the chemical formula for an organic compound is not sufficient information because many isomers (different structural forms with the same chemical formula) can exist.

Organic compounds often exist as mixtures. Because many organic compounds have relatively low boiling points and/or dissolve easily in organic solvents, there exist many methods for separating mixtures into pure constituents that are specific to organic chemistry such as distillation, crystallization and chromatography techniques. Thus, there exist a number of different methods for deducing the structure and/or chemical composition of an organic compound.

There are many more techniques that have specialized applications and within each major analytical technique there are many applications and variations of the general techniques. These include:

- Infrared (IR) and Raman Spectroscopy,
- Atomic spectroscopy,
- Mass Spectrometry,
- Nuclear Magnetic Resonance(NMR),
- X-ray Diffraction,
- UV/Vis Spectrophotometry (for measuring molar concentrations) and
- Chromatography (GC, HPLC, TLC) all of which are employed for detection of various organic compounds and biological samples.

With the advent of Mandelic acid chemistry, obviously need arose to analyze them both qualitatively and quantitatively to perfection. The detection and quantification methods are specific to a particular analyte. Some of these techniques which have been successfully employed for the detection of Mandelic acid are discussed in the following section:

Most of these assessment methods involve the use of chromatography technique. *Chromatography* (from Greek *chroma* "color" and *graphein* "to write") is the collective term for a set of laboratory techniques for the separation of mixtures. A chromatographic method involves a sample being dissolved in a mobile phase (which may be a gas, a liquid or a supercritical fluid). Then through an immobile, immiscible stationary phase the mobile phase is forced. The phases are chosen such that components of the sample have differing solubility in each phase. Component soluble in the stationary phase takes longer to travel than the not very soluble components in the stationary phase but very soluble in the mobile phase. Differences in the partition coefficient of compounds result in differential retention on the stationary phase. Therefore the separation is manifested. As a result of these differences in mobilities, sample components become separated from each other as they travel through the stationary phase.

The technique of chromatography was discovered by a Russian Botanist Tsvet in the mid 19th century. By the first decade of the 20th century, this technique was primarily used for the separation of plant pigments.

Chromatography became developed substantially as a result of the work of Archer John Porter Martin and Richard Laurence Millington Synge during the 1940s and 1950s. They established the principles and basic techniques of partition chromatography, and their work encouraged the rapid development of several types of chromatography method: paper chromatography, gas chromatography and what would become known as high performance liquid chromatography. After that the technology has rapidly advanced. It was seen that the main principles of Tsvet's technique could be applied to many different systems. This resulted in the continually improved performance of chromatography, which allowed the separation of increasingly similar molecules.

Depending on the stationary phase and the mobile phase used, the chromatography techniques used are:

- ➢ Paper chromatography,
- ➢ Thin layer chromatography,
- ➢ Column chromatography,
- ➢ Affinity chromatography,
- ➢ Ion exchange chromatography,
- ➢ Size-exclusion chromatography,
- ➢ Reversed-phase chromatography,
- ➢ Gas chromatography,
- ➢ High-performance liquid chromatography

There are various techniques which are used in combination with chromatographic methods. These techniques have been found to help in better analysis of the substances. These are:

➢ Gas chromatography–mass spectrometry,
➢ Liquid chromatography–mass spectrometry,
➢ Pyrolysis–gas chromatography–mass spectrometry

The chromatography methods used for the detection of Mandelic acid include

➢ Paper chromatography
➢ Thin layer chromatography
➢ High-performance liquid chromatography

6.2 PAPER CHROMATOGRAPHY

Paper chromatography is an analytical chemistry technique for separating and identifying mixtures that are or can be coloured, especially pigments. This can also be used in secondary or primary colours in ink experiments. This method has been largely replaced by thin layer chromatography; however it is still a powerful teaching tool that is used not only for separation but also for assessing the purity and identity of the compound. It uses very small amount of sample and is relatively quick. Two-way paper chromatography, also called two-dimensional chromatography, involves using two solvents and rotating the paper 90° in between the solvent runs. It is used for separating amino acids.

The principal which is based on the substances to be analysed is distributed between a stationary phase (a piece of high quality filter paper) and a mobile phase (a slovent that travels up the stationary phase, carrying the samples with it). Separation of samples depends on the adsorption of compounds on the stationary phase versus their solubility in the mobile phase. Once the analyte is separated, the R_f value can be calculated which is defined as the ratio of the distance travelled by the substance to the distance travelled by the solvent. If R_f value of a solution is zero, the solute remains in the stationary phase and thus it is immobile. If R_f value = 1 then the so lute has no affinity for the stationary phase and travels with the solvent front.

6.3 THIN LAYER CHROMATOGRAPHY

Thin layer chromatography (TLC) is a chromatography technique used to separate mixtures which is performed on a sheet of glass, plastic, or aluminium foil coated with a thin layer of adsorbent material, usually silica gel, aluminium oxide, or cellulose (blotter paper). This layer of adsorbent is known as the stationary phase.

After the sample has been applied on the plate, a solvent or solvent mixture (known as the mobile phase) is drawn up the plate via capillary action. Because different analytes ascend the TLC plate at different rates, separation is achieved.

The principal for TLC is similar to paper chromatography. The advantage over paper chromatography is that of faster runs, better separations, and the choice between different stationary phases. Because of its simplicity and speed, TLC is often used for monitoring chemical reactions and for the qualitative analysis of reaction products. Separation of compounds is based on the competition of the solute and the mobile phase for binding places on the stationary phase. For instance, if normal phase silica gel is used as the stationary phase it can be considered polar. Given two compounds which differ in polarity, the more polar compound has a stronger interaction with the silica and is therefore more capable to dispel the mobile phase from the binding places. Consequently, the less polar compound moves higher up the plate (resulting in a higher Rf value). If the mobile phase is changed to a more polar solvent or mixture of solvents, it is more capable of dispelling solutes from the silica binding places and all compounds on the TLC plate will move higher up the plate.

Table 6.1. Rf value for standard Mandelic acid

Compound	Rf value
Mandelic acid	0.20
Mandelonitrile	0.46

For detection of Mandelic acid solution (0.1 g dissolved in 5 ml methanol) usually Silica gel F254 plate is used, which is activated 110° C for 20-30 mins. The most suitable mobile phase for Mandelic acid is 90: 25: 4; Toluene: Dioxane: Acetic Acid. The separated molecules are observed under UV light in UV chamber.

6.4 HIGH-PERFORMANCE LIQUID CHROMATOGRAPHY

High-performance liquid chromatography (HPLC) is a form of liquid chromatography to separate compounds that are dissolved in solution. HPLC instruments consist of a reservoir of mobile phase, a pump, an injector, a separation column, and a detector. Compounds are separated by injecting a plug of the sample mixture onto the column. The different components in - pass through the column at different rates due to differences in their partitioning behaviour between the mobile liquid phase and the stationary phase.

The sample to be tested is added in small volumes, into the stream of mobile phase. Solvents must be degassed to eliminate formation of bubbles. The pumps provide a steady high pressure with no pulsating and can be programmed to vary the composition of the solvent during the course of the separation. The solution moved through the column is slowed by specific chemical or physical interactions with the stationary phase present within the column. The velocity of the solution depends on the nature of the sample and on the compositions of the stationary phase i.e. the column. The time at which a specific sample elutes is called the retention time; the retention time under particular conditions is considered an identifying characteristic of a given sample. The use of smaller particle size column packing (which creates higher backpressure) increases the linear velocity allowing the components less time to diffuse within the column, improving the chromatogram resolution. Common solvents used include any miscible combination of water or various organic liquids (the most common are methanol and acetonitrile).

The choice of solvents, additives and gradient depend on the nature of the column and sample. Often a series of tests are performed on the sample together with a number of trial runs in order to find the HPLC method which gives the best peak separation.

The detection of Mandelic acid can be carried out by HPLC using a C_{18} (Octyl decylsilinazised) column. The best combination of the mobile phase was found to be 0.01 M Phosphoric acid: Acetonitrile: Methanol in the ratio of 70:27:3. The flow rate was maintained at 1.0ml/ min. A concentration of 1mg/ml OF standard Mandelic acid was used. The detection was carried out at 240 nm.

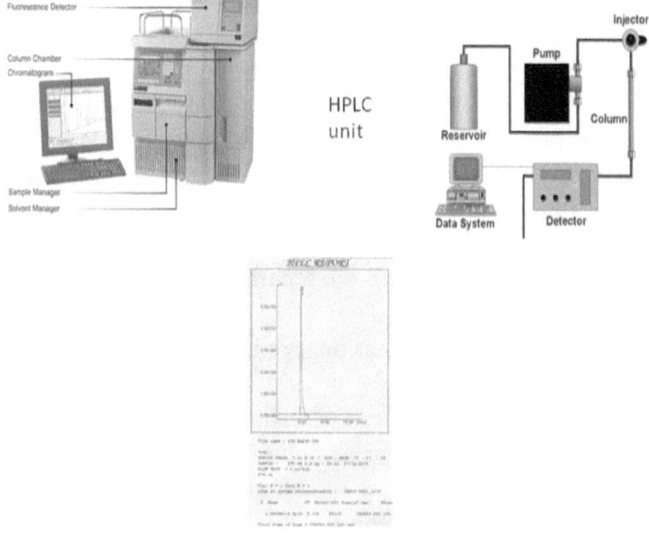

Standard HPLC graph for mandelic acid

Fig 6.1 HPLC unit and HPLC graph for Mandelic Acid

6.5 GAS CHROMATOGRAPHY

Gas Chromatography (GC) is a commonly used analytical technique in many research and industrial laboratories. A broad variety of samples can be analyzed as long as the compounds are sufficiently thermal stable and volatile enough.

Like for all other chromatographic techniques, a mobile and a stationary phase are required. Here the mobile phase (carrier gas) is an inert gas (e.g. Helium, Argon, Nitrogen etc.) and the stationary phase consists of a packed column where the packing or solid support itself acts as stationary phase, or is coated with the liquid stationary phase (high boiling polymer). More commonly used in many instruments are capillary columns, where the stationary phase coats the walls of a small – diameter tube directly (e.g. 0.25 mm film in a 0.32 mm tube)

The main reasons why different compounds can be separated this way is the interaction of the compound with the stationary phase (like dissolves like rule). The stronger the interaction the longer the compound remains attached to the stationary phase, and the more time it takes to go through the column (loner retention time).

Factors that influence the separation are:-

1. Polarity of the stationary phase
    Polar compounds interact strongly with a polar stationary phase, hence have a longer retention time that non-polar columns. Chiral stationary phases based on amino acid derivatives, cyclodextrins, chiral silanes, etc are capable to separate enantiomers, because one form is slightly stronger bonded than the other one, often due to steric effects.
2. Temperature
    The higher the temperature, the more of the compound is in the gas phase. It does interact less with the stationary phase, hence the retention time is shorter, but the quality of separation deteriorates.
3. Carrier gas flow
    If the carrier gas flow is high, the molecules do not have a chance to interact with the stationary phase. The result is the same as the above.
4. Column length - The longer the column, the better is the separation. The trade-off is that the retention time increases proportionally to the column length. There is also a significant broadening of peaks observed, because of increased back diffusion inside the column.

5. Amount of material injected: If too much of the sample is injected, the peaks show a significant tailing, which causes a poorer separation.

6. High temperatures and high flow rates decrease the retention time, but also deteriorates the quality of the separation

Many GC instruments are coupled with a mass spectrometer. GC is used to separate the compounds and the mass spectrometer is used to identify them by assessing their fragmentation pattern. In addition to mass spectrometer (GC/MS), other detectors that are commonly used are

- Flame Ionization Detector (FID)
- Thermal Conductivity Detector (TCD)
- Electron Capture Detector (ECD)

A sensitive and stereospecific GC method was developed for the analysis of R- and S- enantiomers of Mandelic acid (MA) in urine, using a chiral CP Chirasil-Dex-CB column. The enantiomers of MA were derivatised with isopropanol into their corresponding isopropyl esters and determined either directly with flame ionization detection (FID) or after subsequent derivatisation of a hydroxy group with pentafluoropropionic anhydride with electron capture detection (ECD). Both derivatisation steps proceeded with negligible inversion of enantiomers (<1%). The limit of detection of the FID determination was 8 and 5mg/l for R-MA and S-MA, respectively and of the ECD determination 1mg/l for both enantiomers. Repeatability (within day precision) and reproducibility (day to day precision) was for both enantiomers below 7.5% for the FID and below 5.8% for the ECD analysis.

Zougagh *et al.* (2006) have proposed the use of a supercritical carbon dioxide reaction medium for the determination of the (R) - and (S) - enantiomers of Mandelic acid. The process involves a previous derivatization step under supercritical conditions by which the carboxyl group is esterified with methanol that is followed by acylation of the hydroxyl group in methyl Mandelic acid with pentafluoropropionic anhydride in the absence of a catalyst. These derivatization steps cause no enantiomeric inversion. The Mandelic acids where derivatization was done using the method of Kezic *et al.* (2000). The derivatized enantiomers are extracted and quantified by gas chromatography. A BETA DEX 225 capillary column allows the separation of (R)-Mandelic acid and (S)-Mandelic acid as pentafluoropropionyl methyl esters. The overall method was used to determine both enantiomers in urine samples.

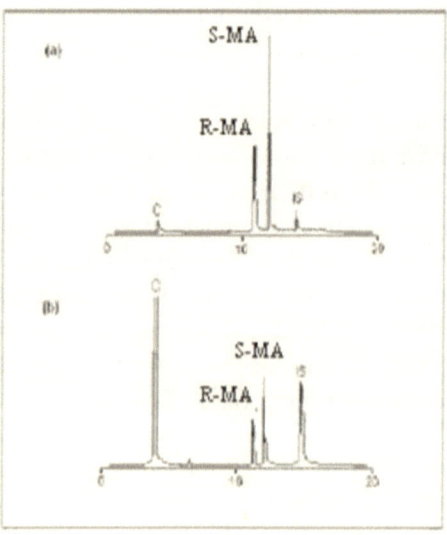

Figure 6.2 - Gas chromatogram for (R)-MA and (S)-MA after esterification and subsequent acylation with PFPA of a urine sample spiked with 100 μg mL–1 concentrations of (R)-MA and (S)-MA, using the proposed SFE method (*a*) and an LLEmethod (*b*). IS internal standard; C unknown peak
(*Courtsey Valcarcel et al*)

6.6 MASS SPECTROMETRY

Mass spectrometry (MS) is used for the determining the elemental composition of a sample or molecule as well as for elucidating the chemical structures of molecules by ionizing the compounds (to generate charged molecules or molecule fragments) and measuring their mass-to-charge ratios.

In a typical MS procedure:

1. A sample is loaded onto the MS instrument, and undergoes vaporization.
2. The components of the sample are ionized by one of a variety of methods (e.g., by impacting them with an electron beam), which results in the formation of charged particles (ions)
3. The positive ions are then accelerated by an electric field
4. Computation of the mass-to-charge ratio (m/z) of the particles based on the details of motion of the ions as they transit through electromagnetic fields and
5. Detection of the ions, which in step 4 were sorted according to m/z.

MS instruments consist of three modules:

1. An ion source, which converts gas phase sample molecules into ions,
2. A mass analyzer, which sorts the ions by their masses by applying electromagnetic fields; and
3. A detector, which measures the value of an indicator quantity and thus provides data for calculating the abundances of each ion present.

The technique has both qualitative and quantitative uses. These include

> ➢ Identifying unknown compounds,
> ➢ Determining the isotopic composition of elements in a molecule
> ➢ Determining the structure of a compound by observing its fragmentation
> ➢ Quantifying the amount of a compound in a sample
> ➢ Studying the fundamentals of gas phase ion chemistry

MS is now in very common use in analytical laboratories that study physical, chemical, or biological properties of a great variety of compounds.

A common combination is Gas Chromatography-mass spectrometry (GC/MS or GC-MS). In this technique, a Gas Chromatography is used to separate different compounds. This stream of separated compounds is fed online into the ion source, a metallic filament to which voltage is applied. This filament emits electrons which ionize the compounds. The ions can then further fragment, yielding predictable patterns. Intact ions and fragments pass into the mass spectrometer's analyzer and are eventually detected.

6.7 NUCLEAR MAGNETIC RESONANCE SPECTROSCOPY

Nuclear Magnetic Resonance (NMR) Spectroscopy is the name given to a technique which exploits the magnetic properties of certain nuclei. The most important applications for the organic chemist are proton NMR and Carbon-13 NMR spectroscopy. In principle, NMR is applicable to any nucleus possessing spin.

Many types of information can be obtained from an NMR spectrum. Much like using infrared spectroscopy (IR) to identify functional groups, analysis of a NMR spectrum provides information on the number and type of chemical entities in a molecule. It is seen that NMR provides much more information than IR.

The impact of NMR spectroscopy on the natural sciences has been substantial.

NMR spectroscopy can be used

> To study mixtures of analytes,
> To understand dynamic affects such as change in temperature and reaction mechanisms,
> Is an invaluable tool in understanding protein and nucleic acid structure and function
> It can be applied to a wide variety of samples, both in the solution and the solid state.

Scapolla *et al.* (2010) used a mixture of triphenylphosphine and 2,2'-dipyridyl disulfide (Mukaiyama reagent), an oxidation-reduction condensation reagents, as coupling reagents for the formation of amide bond. In their reaction of the DL-VanilMandelic (VMA) and DL-Mandelic Acid (MA) with (S)- Methylbenzylamine in presence of the Mukaiyama reagent over the expected diastereomeric amides some fluorescents by-products were formed, which were characterized by Nuclear Magnetic Resonance (1H-NMR, 13C-NMR, COSY, HSQC, HMBC), Mass Spectrometry (High Resolution MS Tandem MS/MS) and Infrared Spectroscopy for structure identification which they suggested to be a mesoionic structure like compounds as "thioisomunchnones" (1,3-thiazolium-4-olates). Hence, they derivatized MA as thioisomunchnone for its detection and quantification in liquid chromatography coupled with a fluorescence detector.

6.8 CONCLUSION

Various analytical methods for identification and separation of Mandelic acid are discussed in this chapter. Simple procedures like paper chromatography as well as complex analytical procedure using NMR; all can be employed to assess the Mandelic acid. This chapter gives an insight that Mandelic acid detection can be carried out even in less equipped labs with minimal facilities.

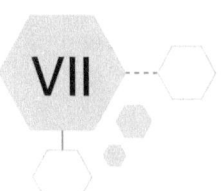

VII

BIOCONVERSION OF MANDELONITRILE TO MANDELIC ACID

7.1 INTRODUCTION

The process of chemical and physical change goes on continually in the living organisms. Build-up of new tissue, replacement of old tissue, conversion of food

> *Only two things are infinite, the universe and human stupidity, and I'm not sure about the former*
> *- Albert Einstein*

to energy, disposal of waste materials, reproduction-all these activities that is characterized as "life" takes place in the face of an apparent paradox. The phenomenon of catalysis that involves enzymes as *"agents of life"* makes possible the biochemical reactions which are necessary for all life processes, thus showing a unique property of welding the molecules in the cell by enzymatic action. These activities involve biochemical reactions that do not take place spontaneously.

This uniqueness of enzymes is widely exploited for bioconversion processes thus replacing the traditional chemical ways.

The main objective of using bioconversion process here is to use biosystem as source of enzyme to produce Mandelic acid.

Commercially Mandelic acid is prepared by a chemical method from Benzaldehyde as precursor. This reaction is not specific and results in the formation of both R (-) and S (+) Mandelic acid, which are enantiomers of each other. To acquire only R (-) Mandelic acid (which has a great demand in skin care industry), racemization of the S (-) Mandelic acid is carried out to form R (-) Mandelic acid. Nitrilase enzyme is specific for the production of R (-) Mandelic acid. The only problem is that Nitrilases are very expensive and not easily available.

Nitrilases are commercially used for conversion of Mandelonitrile to Mandelic acid, nicotinic acid and the detoxification of cyanide waste. Apart from chemically procured Nitrilases; efforts have been successfully made to use microbes and plants as source of Nitrilases.

Having colonized every niche and habitat, bacteria, archea and plants have evolved enzymatic solutions for wide range of metabolic biochemical transformations. They have contributed immensely in White Biotechnology wherein they are exploited for production processes like fermentations and bioconversions. This approach being environmentally safe has been accepted as Green Chemistry, wherein the design of chemical products and processes reduce or eliminate the use and generation of hazardous substances, thus reducing and preventing pollution at its source, it mainly focuses on minimizing the hazard and maximizing the efficiency of the chemical choice.

In this chapter we are presenting efforts made using microbes and plants for the production of Mandelic acid.

7.2 NITRILASE CATALYZED CONVERSION OF MANDELONITRILE TO MANDELIC ACID BY MICROBES

Several microorganisms have been reported to hydrolyse both aliphatic and aromatic compounds. In bacteria, hydration of nitriles proceeds by two different enzymatic pathways:

(1) A Hydratase/Amidase enzyme system which catalyses the reaction to the corresponding carboxylic acid and ammonia in a two-step reaction via a free amide intermediate, or

(2) A Nitrilase enzyme which catalyses the reaction in one step.

In addition to these hydration reactions, many naturally occurring Nitrilases and reduction of nitriles by these nitrogenase enzymes has been described in many instances. Indeed, most nitrile hydrolysing bacteria described in detail have been enriched and finally isolated using media containing synthetic nitrile as carbon and/or nitrogen sources. Many synthetic nitrile compounds are used in industrial operations. The widespread usage of nitrile has created considerable commercial interest in nitrile-hydrating enzymes, both as biocatalysts for chemical syntheses and as a means of biological detoxification of nitrile-containing waste. Industrial usage of nitriles may result in discharge of synthetic nitriles into marine, freshwater and soil environments, where they may be detoxified by natural populations of nitrile-hydrolysing microorganisms. All nitrile-hydrolysing bacteria described in the literature appear to have been isolated from soils or freshwater environment.

A number of naturally occurring nitriles have been identified. Since most of these are not commercially available they have not been tested as possible substrates for nitrile hydrolysing enzymes. So far only a few biogenic nitriles (e.g. ricinine and mandelonitrile) have been reported to be hydrolysed by microorganisms. Inorganic cyanide (HCN/CN-) is produced in vast amounts by natural synthesis and is, therefore, of widespread occurrence in nature. However, known Nitrile hydratase and Nitrilases are unable to hydrolyse cyanide. Another explanation for the presence of nitrile-hydrating enzymes in bacteria could be that these enzymes have an unspecific enzymatic reaction.

Microbes metabolize cyanosugars, cyanohormones, plant auxins, and other organocyanide compounds produced for defense or storage by bacteria and eukaryotes. Also, some microorganisms convert nitrile compounds to the corresponding acids by the action of Nitrilase or of Nitrile hydratase in combination with Amidases. Nitrilase producing microorganisms are *Alcaligenes faecalis, Pseudomonas vesicularis, Candida guilliermondii, Rodococus sp., Corynebacterium sp., Mycobacterium Sp.,* and *Acinetobacter sp.,* R-Mandelic acid was produced from racemic Mandelonitrile using free and immobilized cells of *Pseudomonas putida* MTCC 5110 harboring a stereo-selective Nitrilase.

Researchers in China conducted the Nitrilase-mediated manufacture of R-(–)- Mandelic acid. He *et al.* (2007; 2010) screened a microbial strain identified as *Alcaligenes* sp. ECU0401 harbouring a stereo-selective Nitrilase for the kinetic resolution of racemic Mandelonitrile to (R)-(–)-Mandelic acid with an enantiomeric excess of >99.9%. This convenient and practical approach to producing (R)-(–)-Mandelic acid was developed into commercial application by Mitsubishi Rayon.

R-(-) Mandelic acid was produced from racemic Mandelonitrile by *Alcaligenes faecalis* ATCC 8750, using Ammonium acetate or L-glutamic acid as the carbon source and n-butyronitrile as the inducer in the culture medium. R-(-) Mandelic acid was produced from racemic Mandelonitrile in a yield of 91% while no S-Mandelonitrile was left. R-(-) Mandelic acid was also produced when benzaldehyde along with HCN was used as the substrate. Out of the screened Nitrilases obtained from various microorganisms, 20 Nitrilases have been found to convert Mandelonitrile to Mandelic acid with >90 % enantiomeric excess (ee). One of these tested enzymes was found to produce (R) Mandelic acid at high rates (i.e. within 10 min of reaction at 0.12 mg/ml enzymes) with high enantio-selectivity (98 %ee) and in high yield (86 % isolated on 1.0 g scale). This enzyme was shown to be equally effective on a range of ortho-, meta-, and para- substituted Mandelic derivatives and analogues.

Nitrilase of *Alcaligenes faecalis* JM3 and *Alcaligenes faecalis* ATCC 8750 are inducible, while the Nitrilase of *Alcaligenes faecalis* ECU0401 is a constitutive enzyme. A Nitrilase gene blr3397 from *Bradyrhizobium japonicum* USDA110 was cloned and over-expressed in *E. coli*, which showed higher activity towards the hydrolysis of aliphatic nitriles than that for the aromatic counterparts. But the e.e. value and conversion of (S)-(-)-Mandelic acid were 36 and 44%, respectively. Kaul *et al* (2004) reported that the e.e. values of (R)-(-)-Mandelic acids were very high at the start of the reaction with three bacterial isolates, but dropped as the reaction proceeds and more product formation occurred. In our case, the e.e. value of (R)-(-)-Mandelic acid was kept above 99.9%. However, the yield of

(R)-(-)- Mandelic acid was low (B55.0%) in our trials. The objective of the present study was to enhance the catalytic efficiency of *Alcaligenes faecalis* ECU0401.

Nagasawa *et al* (2000) have also successfully produced R-Mandelic acid from racemic mandelonitrile using free and immobilized cells of *Pseudomonas putida* MTCC 5110 harboring a stereo selective Nitrilase. Similar work, done by Kaul *et al* (2004) revealed that the Nitrilase from *Rhodococcus rhodochrous* J1, which catalyses the hydrolytic cleavage of the C-N triple bond in nitrile to form acid and ammonium, catalyses hydrolysis of amide to acid and ammonium stoichiometrically. The existence of common tetrahedral intermediate in the Nitrilase reaction involving nitrile or amide as a substrate has been noted.

Biotransformation was performed with recombinant *Escherichia coli* strains that heterologously expressed Nitrilase activities originating from *Pseudomonas, Rhodococcos* or *Synechocystis* strains. The attempted conversion of the 2-acetoxynitriles to almost stoichiometric amounts of the corresponding 2-acetoxycarboxylic acids was achieved by using either a recombinant *E. coli* strain that highly over expressed the Nitrilase gene from *Pseudomonas* or purified enzyme derived from this strain (Rustler *et al* 2007). The gene encoding an enantio-selective aryl-acetonitrilase was identified on a 3.8kb DNA fragment from the genomic DNA of *Pseudomonas fluorescence* EBC 191. The gene was isolated, sequenced and cloned into the l-rhamnose inducible expression vector pJOE2775. The Nitrilase was produced in large quantities and purified as a histidine –tagged enzyme from crude extracts of l-rhamnose induced cells of *Escherichia coli* JM109. The specific activity of the recombinant Nitrilase was similar to those of the native enzyme from *P. fluorescence* EBC191 (Banerjee *et al* 2002).the enzyme hydrolyzed various phenyl-acetonitriles with different substituent in the 2-position and also heterocyclic and bicylcic arylacetonitriles to the corresponding carboxylic acids. The conversion of most arylacetonitriles was accompanied by the formation of different amounts of amides as by products. The relative amounts of amides formed from different nitriles increased with an increasing negative inductive effect of the substituents in the 2-position. the acids and amides that were formed from chiral nitriles demonstrated in most cases opposite enantiomeric excesses. Thus Mandelonitrile was converted by the Nitrilase preferentially to R-Mandelic acid and S-Mandelic acid amide. The Nitrilase gene is physically linked in the genome of P. *fluorescence* with genes encoding the degradative pathway for Mandelic acid; suggesting a natural function of the nitrilase in the degradation of Mandelonitrile or similar naturally occurring hydroxynitriles.

A process for the preparation of D(-) Mandelic acid by the acid hydrolysis of the corresponding carbamate in an aqueous environment at a temperature of between 40°C to 100° C and at a pH of between 1 and 3.5 is already patented. The invention also relates to a process for the enzymatic hydrolysis of racemic 5-substituted 2, 4-oxazolidinediones to give only one of the two possible optical isomers, i.e., the D-carbamyl-α-hydroxy acid. The free D-.α-hydroxy acid can be obtained from the optically active carbamyl derivative by simple hydrolysis. Of particular interest is the case in which the D-α-hydroxy acid is D (-) Mandelic acid. The enzymatic activity required for preparing the carbamyl derivative of D (-) Mandelic acid has been found both in homogenized veal liver and in a series of microorganisms, including *Agrobacterium radiobacter, Bacillus brevis, Bacillus stearothermophilus, Pseudomonas sp., Pseudomonas desmolytica, Pseudomonas fluorescens, and Pseudomonas putida.*

Nitrile-hydrolysing bacteria: A number of physiologically different nitrile-hydrolysing bacteria were isolated from coastal marine sediments in Denmark by enrichment culture and were identified as families belonging to *Corynebacteriaceae, Mycobacteriaceae (Rhodococcus erythropolis)* and *Nocardiaceae (Nocardia calcarea).* One strain, BLI, identified as *Rhodococcus erythropolis,* grew on acetonitrile as sole carbon and nitrogen source in a defined medium. Intact cells of *R. erythropolis* BLI could hydrolyse a large variety of saturated and unsaturated aliphatic nitriles to their corresponding acids. Benzonitrile and benzylcyanide could not be hydrolysed, whereas some aromatic compounds containing a -CN group attached to a C, or C4 aliphatic side chain were accepted as substrates. The substrate spectrum of *R. erythropolis* BLI was thus markedly different from those of other Gram positive nitrile-hydrolysing bacteria isolated from non-marine environments. Nitrile hydrolysis during growth and in resting cell suspensions usually occurred without intermediate accumulation of amide outside the cells. Detailed studies showed that nitrile hydrolysis by strain BLI was due to a nitrile Hydratase/Amidase enzyme system. Nitrile Hydratase activity was found to be inducible whereas Amidase activity was constitutive. The Amidase activity of cells could, however, be enhanced many fold by growth in media containing acetamide or acetonitrile. In most cases amides were hydrolysed at a much higher rate than the corresponding nitriles, which explained why amides were rarely detected in the surrounding medium during nitrile hydrolysis. *R.* erythropolis BLI exhibited the highest tolerance towards acetonitrile ever reported for a nitrile-hydrolysing bacterium, as demonstrated by its ability to grow exponentially in the presence of 900 mM acetonitrile.

(R)- Mandelic acid was produced from racemic mandelonitrile using free and immobilized cells of *Pseudomonas putida* MTCC 5110 harboring a stereo-selective Nitrilase (Kabaivanova *et al* 2005). In addition to the optimization of culture conditions and medium components, an inducer feeding approach is suggested to achieve enhanced enzyme production and therefore higher degree of conversion of mandelonitrile. Efficient biocatalyst recycling was achieved as a result of immobilization with immobilized cells exhibiting 88% conversion even after 20 batch recycles. A fed batch reaction set up on a preparative scale produced 1.95 g of (R)-(-)-Mandelic acid with an enantiomer excess of 98.8%.

The stability and reusability aspect of Nitrilase from *Alcaligenes faecalis* for the production of (R)-(-) - Mandelic acid was seen by entrapment of the cells. Four entrapment matrixes were screened to search for a suitable support, and alginate was found to have significant process advantages over its other counterparts (Zhu *et al* 2007). Efficient reusability of the biocatalyst upto 35 batches was achieved by immobilization as compared to 9 batches for free cells and cross-linking extended it further to 40 batches. Finally, synthetic utility of the immobilized biocatalyst was demonstrated on a preparative scale to produce 640 g of (R)-(-)-Mandelic acid with 97% enantiomeric excess.

Biocatalytic enantio-selective hydrolysis of beta-hydroxy nitriles to corresponding (S)- enriched beta-hydroxyl carboxylic acids has been achieved for the first time by an isoloted Nitrilase bll6402 from *Bradyrhizobium japonicum* USDA110. This offers a new "green" approach to optically pure beta-hydroxy nitriles and beta-hydroxy carboxylic acids. The observed remote stereo-recognition is surprising because this Nitrilase showed no enantio-selectivity for the hydrolysis of alpha- hydroxy nitriles such as mandelonitrile (Kamila *et al* 2006; Quing *et al* 2007).

7.2.1 OUR EFFORTS USING MICROBES

Using Alcaligenes faecalis and Acinetobacter sp: Our group worked on the production of R- (-)-Mandelic acid from

Mandelonitrile; using 5 different microbes i.e. *Alcaligenes faecalis* ATCC 8750 (NCIM No.-2262); *Alcaligenes faecalis* ATCC 8750 (NCIM No.-2105); *Acinetobacter. sp.* (NCIM No. - 5083); *Acinetobacter calcoaceticus* ATCC 14987 (NCIM No.-2890) and *Acinetobacter calcoaceticus* ATCC 11171 (NCIM No.-2886)

Maximum conversion rate of 7.68% was obtained by using *A. faecalis ATCC 8750*. The bacterial cells were grown aerobically at 28° C for 2 days in a nutrient rich medium (20 g Peptic digest, 2 g D-Mannitol, 2 g NaCl, 1.5 g $MgSO_4.7H_2O$, 1 g of K_2HPO_4, 1 g L-Arginine, 1 g NaNO, 1g Yeast extract, 0.35 g $NaNO_3$, 15 ml of 5g/ml filter sterilized Glucose solution/L) containing 3 g n-butyronitrile which acts as an enzyme inducer. The pH of the medium was maintained at 6.7 ± 0.2.

Cells from the log phase of growth were harvested and centrifuged at 3500 rpm for 10 min to obtain a cell pellet, which was re-suspended in 5 ml of 0.1 M Potassium phosphate buffer (pH-8.0) and used as source of enzyme Nitrilase. Activity of Nitrilase was assessed using conversion of acetonitrile to acetic acid and detected by paper chromatography. The mobile phase for paper chromatography was 10:1:1 iso-propanol: ammonia: water and indicator was 0.4% bromo-cresol purple in ethanol.

Bioconversion of Mandelonitrile to Mandelic acid

To the 5 ml of re-suspended cells 2 ml of Mandelonitrile was added. The samples were incubated at three different temperatures i.e. at 26, 28, 30 and 32° C for 24 hrs.

After the incubation, presence and quantification of Mandelic acid was done by Thin Layer Chromatography and High Performance Liquid Chromatography respectively.

For Thin Layer Chromatography, activated Silica gel F254 plates were used. A Mobile phase of Toluene: Dioxane: Acetic acid (90: 25: 4) was used. 0.1 gm of the sample was dissolved in 5 ml of methanol. Chromatogram was run; the plates were dried and observed under UV light.

For High Performance Liquid Chromatography, C_{18} (Octyl decylsilinazised) column was used. The mobile phase had 0.01 M Phosphoric acid: Acetonitrile: Methanol in the ratio of 70:27:3. 20 µl of the sample solution was injected. A UV detector of 240 nm was used. A flow rate of1.0 ml/min was constantly maintained. 1 mg/ml Mandelic acid was used as the standard reference.

Alcaligenes faecalis ATCC 8750 and *Acinetobacter sp* used were similar to that tried by Yamamoto *et al* (1991). However, the best temperature for conversion of Mandelonitrile to Mandelic acid was not the same as recorded by Yamamoto *et al*. They could get maximum conversion at 32°C where as in the present work it was 28°C for all the 5 strains of microbes.

Alcaligenes faecalis ATCC 8750 is quite famous for the effective hydrolysis of Mandelonitrile to (R)-(–)-Mandelic acid. Interestingly, enantiomeric excess of (R)-(–)-Mandelic acid formed from Mandelonitrile was 100% and the yield attained was approximately 91%. Spontaneous racemisation of S-Mandelonitrile because of the chemical equilibrium accounted for the high yield. The observation was inconsistent with other enantio-selective bioprocess where the yield attained was 50% at most.

After successful purification and characterization of the Nitrilase, immobilization of the Nitrilase was tried. In particular, the immobilized Nitrilase is useful for the production of hydroxy analogues of methionine derivatives that could have an interest in cattle feeding and in the transformation of compounds bearing other acid- or base-sensitive groups.

Using Free and Immobilized Cells of *Pseudomonas putida* MTCC 5110 –

For immobilization of the cells 50 ml of cell suspension was mixed with 50 ml of 4 % and 2% sodium alginate separately. The well-mixed solution was drop-wise added in chilled 6% calcium chloride solution using a syringe. The beads obtained were incubated in the calcium chloride solution at 4- 6 ° C for 24 hrs for hardening. The beads were filled in a 100 ml conical flask containing 2 ml of Mandelonitrile. The samples were incubated at 28 °C for 24 h. After the appropriate incubation period, the analysis of Mandelic acid was carried out by TLC and HPLC.

A comparison between the conversion capacities of immobilized and un-immobilized microbial cell suspension was done. *Alcaligenes faecalis* ATCC 8750 (NCIM No.2262) showed better conversion than the other microorganisms. So far as immobilization by alginate was concerned, it could not increase the Nitrilase activity, rather it was inhibitory.

(R)-Mandelic acid was produced from racemic Mandelonitrile using free and immobilized cells of *Pseudomonas putida* MTCC 5110 harboring a stereo-selective Nitrilase. In addition to the optimization of culture conditions and medium components, an inducer feeding approach to achieve enhanced enzyme production and therefore higher degree of conversion of Mandelonitrile was tried. The Nitrilase expression was also authenticated by the sodium dodecyl phosphate-polyacrylamide gel electrophoresis analysis. *P. putida* MTCC 5110 cells were further immobilized in calcium alginate, and the immobilized biocatalyst preparation was used for the enantio-selective hydrolysis of Mandelonitrile. The immobilized system was characterized based on the Thiele modulus (phi). Efficient biocatalyst recycling was achieved as a result of immobilization with immobilized cells exhibiting 88% conversion even after 20 batch recycles. Finally, a fed batch reaction was set up on a preparative scale to produce 1.95 g of (R)-(-)-Mandelic acid with an enantiomeric excess of 98.8%.

The stability and reusability aspect of Nitrilase from *Alcaligenes faecalis* for the production of (R)-(-)-Mandelic acid was studied by entrapment of the cells in alginate. Efficient reusability of the biocatalyst up to 35 batches was carried out by immobilization as compared to 9 batches for free cells, and cross-linking extended it further to 40 batches. Finally, synthetic utility of the immobilized biocatalyst was demonstrated on a preparative scale to produce 640 g of (R)-(-)-Mandelic acid with 97% enantiomeric excess (ee).

7.3 NITRILASE CATALYZED CONVERSION OF MANDELONITRILE TO MANDELIC ACID BY FUNGI

A variety of fungal species degrade cyanide through the action of Cyanide hydratase, a specialized subset of Nitrilases which hydrolyse cyanide to formamide. Basile and Willson reported work on two previously unknown and uncharacterized Cyanide hydratases from *Neurospora crassa* and *Aspergillus nidulans*. Later recombinant forms of four cyanide hydratases from from *N.crassa, A.nidulans, Giberella oryza,* and *Gloeocercospora sorghi* were prepared after their genes were cloned with N-terminal hexa-histidine purification tags, expressed in *Escherichia coli,* and purified using immobilized metal affinity chromatography.

These enzymes were compared according to their relative specific activity, pH activity profiles, thermal stability, and ability to remediate cyanide contaminating waste water from silver and copper electroplating baths. It was observed that although all four were similar, the *N.*

crassa Cyanide hydratase (CHT) has the greatest thermal stability and widest pH range of >50 % activity. *N. crassa* also demonstrated the highest rate of cyanide degradation in the presence of both heavy metals. The CHT of *A. nidulans* having the highest reaction rate of four fungal Nitrilases was evaluated. *2-Cyanopyridine* and *Valeronitrile* are powerful Nitrilase inducers in fungi namely, Aspergillus sp., Fusarium sp. and Penicillium sp.*Several other fungi also have this enzyme; but they do not excrete it into the culture medium.*

7.4 NITRILASE CATALYZED CONVERSION OF MANDELONITRILE TO MANDELIC ACID BY PLANTS

Pace and Brenner (2001) have demonstrated that Nitrilase and Nitrile Hydratase (in combination with Amidases) are the two enzymes, which are mainly involved in the hydrolysis of nitrile compounds in plants, animals and microorganisms. They carry out either one step or two-step enzymatic reaction. Nitrilase catalyzes the mild hydrolytic conversion of organo-nitriles directly to the corresponding carboxylic acids.

The biosynthesis of IAA in plants can proceed via different pathways, which involve indole-3-acetonitrile (IAN) as an intermediate. The enzyme Nitrilase, which converts IAN to the auxin indole-3-acetic acid (IAA), is considered to be a key enzyme in this biosynthetic pathway. Bioconversion or biotransformation is mostly being tried using microbes as source of enzyme (Yamamoto *et al* 1991). However, plants have also been considered as source of Nitrilase for bioconversion to form Mandelic acid. The potential of plant Nitrilases to convert indole-3-acetonitrile into the plant growth hormone indole-3-acetic acid has earned them the interim title of "key enzyme in auxin biosynthesis".

Recent work on plant Nitrilases has shown them to be involved in the process of cyanide detoxification, in the catabolism of cyanogenic glycosides and presumably in the catabolism of glucosinolates. All plants possess at least one Nitrilase that is homologous to the Nitrilase 4 isoform of *Arabidopsis thaliana*. The general function of these Nitrilases lies in the process of cyanide detoxification, in which they convert the intermediate detoxification product ß-cyanoalanine into asparagine, aspartic acid and ammonia. Cyanide is a metabolic by-product in biosynthesis of the plant hormone ethylene, but it may also be released from cyanogenic glycosides, which are present in a large number of plants.

In *Sorghum bicolor*, an additional Nitrilase isoform has been identified, which can directly use a catabolic intermediate of the cyanogenic glycoside, thus enabling the plant to metabolize its cyanogenic glycoside without releasing cyanide. In the *Brassicacea* family, a family of Nitrilases has evolved, the members of which are able to hydrolyze catabolic products of glucosinolates, the predominant secondary metabolites of these plants. Thus, the general theme of Nitrilase function in plants is detoxification and nitrogen recycling, since the valuable nitrogen of the nitrile group is recovered in the useful metabolites asparagine or ammonia. Taken together, a picture emerges in which plant Nitrilases have versatile functions in plant metabolism, whereas their importance for auxin biosynthesis seems to be minor.

This enzyme is not common in the plant kingdom, for of 29 plant (from 21 families) tested, 19 showed no activity and only members of the *Gramineae* (grasses), *Cruciferae* (cabbage group and radish), and *Musaceae* (banana family) were clearly active. The enzyme has been partially purified from barley leaves. (Thimann and Mahadevan, 1963).

Nitrilases of plants are divided into two groups i.e. NIT1 - which is restricted to *Brassicaceae* having broad substrate specificity and are involved in glucosinolate catabolism and NIT4 - which is widely distributed in the plant kingdom having high substrate specificity for b-cyanoalanine and are involved in cyanide detoxification. Benzaldehyde (a precursor for Mandelic acid synthesis) has been extracted from number of plants such as bitter almond, apricot, cheery, laurel leaves, peach seeds. Moreover, a glycosidic form amygdalin has been reported in some nuts and kernels (Quirino *et al* 1999). Hence, these plants have been accepted as source of Nitrilase activity.

Nitrilase are able to convert indole-3-acetonitrile to IAA in vitro and Nitrilase genes are up-regulated during senescence in *Arabidopsis*. At the same time, free IAA levels increase two-fold in senescing leaves while IAN and conjugated IAA levels drop two-fold.

7.4.1 OUR EFFORTS USING PLANTS

Taking a clue from the findings of Thimann and Mahadevan, our group used Barley leaves from young seedlings, Cabbage and Radish (bought from the local market) as the source of enzyme Nitrilase. This economically eco-friendly method is described below:

Extraction of enzyme from plants: The surface of the plant material was washed with distilled water and cut in to small pieces with a knife.10 gm of the each plant material was taken separately in a pre-cooled mortar as extraction of the enzyme has to be done in cold environment. To the chopped plant material: 5 ml of potassium phosphate buffer (to control the pH of the extract and stabilize the protein), 2 ml of 1×10^{-3} M EDTA (for chelating metal ions), 0.2 gm potassium chloride (to maintain the standard cellular environment), 1 ml of 1×10^{-3} M Cysteine (for preventing oxidation of proteins), 5 ml of 5% Glycerol (to reduce the polarity) and a pinch of phenyl methyl sulfonylurea fluoride (as protease inhibitor) was added. The mixture was grinded using mortar and pestle, filtered through muslin cloth and then centrifuged at 10000 rpm at 4°C for 15 minutes (Boyer's, Enzyme extraction techniques, *Modern Experimental Biochemistry*).

The extract obtained was used as a source of enzyme. Activity of Nitrilase was assessed using conversion of acetonitrile to acetic acid and detected using paper chromatography. The mobile phase was 10:1:1 iso-propanol: ammonia: water and indicator was 0.4% bromo-cresol purple in ethanol.

Bioconversion of Mandelonitrile to Mandelic acid: The extract obtained was used as the source of enzyme. 2 ml of extract was added to 2 ml 0.1 M Potassium phosphate buffer of pH 8.0 in each test tube. To this 2 ml of neutralized Mandelonitrile (with potassium phosphate buffer) was added. The samples were incubated at 28, 30 and 32°C for 24 h.

After the appropriate incubation period, presence of Mandelic acid was analyzed using Thin Layer Chromatography (to detect the presence of Mandelic acid) and High Performance Liquid Chromatography (to quantify the Mandelic acid formation).

Crude plant extracts as the source of enzyme have been tried thus confirmed (Pathak et al 2007) the presence of enzyme Nitrilase in Barley, Radish and Cabbage extracts taken in potassium phosphate buffer, EDTA, potassium chloride, Cysteine, Glycerol and phenyl methyl sulfonylurea fluoride (as protease inhibitor). When this extract was added to Mandelonitrile; a High Performance Liquid Chromatography analysis showed temperature specific conversion of Mandelonitrile to Mandelic acid. It was 32°C for Cabbage and Radish, whereas 28°C for Barley. However the conversion rate was very low. From the data analyzed it was found that Barley proved to be a better source of Nitrilase. Cabbage and Radish were more or less the same. So far as immobilization by alginate is concerned, it could not increase the Nitrilase activity, rather it was inhibitory. Hence, scaling up and improvising all the parameters for immobilization and enzyme activity was recommended. The preliminary results have indicated that plants can be used for conversion of Mandelonitrile to Mandelic acid; though lot of optimization and various parameters need to be studied further.

7.5 CONCLUSION

Current research suggests that Nitrilases play important roles in a range of biological processes. The possible applications of Nitrilase in many processes are being researched. Optimization of parameters of the reactions needs to be improved to establish commercially viable industrial applications. A greater need for screening the new sources of enzymes, improvement in enzyme, substrate structure modification, medium engineering, and variation in process parameters is the demand of the day. New Nitrilases have been obtained by genome mining and the metagenomic approach. Variation in substrate structure proved to be another means to modify the enantio-selectivity and chemo-selectivity of the reactions. The current knowledge of the role of nitriles and Nitrilases in plants and plant associated microorganisms will help in greater understanding of the natural functions of Nitrilases could be applied to benefit both industry and agriculture. Moreover, to increase the utility of Nitrilases it would be a good idea to use it in combination with other enzymes or catalysts without laborious intermediate isolation. Such processes have been developed for the synthesis of (S)-Mandelic acid.

Few words we like to express in conclusion . . . .

The modern-day practitioners often regard the medicinal use of herbs as "quackery;" nothing greater than old-wives tales. There are a growing variety of otherwise conventional medical professionals who acknowledge what Grandmothers knew all along. Natural remedies as a means to keep up good health and cure certain diseases are valid.

Another better way to explain traditional and modern methods of treatment—is explaining the headache. Now-a- days if we get a headache, we tend to take an aspirin. The traditional way was getting a good night sleep instead of consuming unnecessary tablets and numbing the pain.

In the natural world, besides herbs, many greens and fruits, particularly organic, yield health and medicinal benefits. Dried burdock root is used as a blood purifier. It can be the "king" of herbs in treating continual skin issues such as eczema, acne, psoriasis, boils, syphilitic sores, and canker sores. Celery juice is a natural diuretic and helpful for persons with rheumatism or for those who need to lose weight. Cabbage has been proven effective within the fight against duodenal ulcers, and is a good source of calcium for those who must avoid dairy products. Radish is useful for gall-bladder and liver ailments, and spinach *improves the hemoglobin of the blood. Beets are excellent for certain conditions of the liver, and for enhancing blood hemoglobin.*

With the proper knowledge of the human systems and the regulations, the world of medicine changed. The development of modern neurology began in the 16th century with Vesalius, who described the anatomy of the brain. Understanding and diagnosis improved but with little direct benefit to health.

Medicine was revolutionized in the 19th century and beyond by advances in chemistry and laboratory techniques and equipment, old ideas of infectious disease epidemiology were replaced with bacteriology and virology. The 1953 discovery of the structure of DNA by Watson and Crick would open the door to molecular biology and modern genetics.

The Great War spurred the usage of Roentgen's X-ray, and the electrocardiograph, for the monitoring of internal bodily functions. This was followed in the inter-war period by the development of the first anti-bacterial agents such as the sulpha antibiotics. The Second World War saw the introduction of widespread and effective antimicrobial therapy with the development and mass production of penicillin antibiotics, made possible by the pressures of the war and the collaboration of British scientists with the American pharmaceutical industry The 20th century witnessed a shift from a master-apprentice paradigm of teaching of clinical medicine to a more "democratic" system of medical schools. With the advent of the evidence-based medicine and great advances of information technology the process of change is likely to evolve further, with greater development of international projects such as the Human genome project.

It was believed that eating almonds soaked in warm water was supposed to be good for healthy hair, beautiful flawless skin. Little was it known at that time that it contained the secret ingredient in the form of Mandelic acid…. AHA…. that was the reason for its beautifying property.

Archaeologists have found that ancient Egyptians use an inclusion of bitter almonds for daily skin care to improve the elasticity of skin. This was found in a translation of a 3,000-year-old papyrus scroll detailing surgical practices of those days. Decoding this information needed a decade of archaeological travel and research trying to find the correct modern-day translation of the key ingredient by Hemayet, in a chapter entitled "Transforming an Old Man into a Youth." It turned out the word meant "bitter almond," which contains Mandelic acid.

Scientists have long known that Mandelic acid is an alpha hydroxy acid and that its molecule is larger and has a lower penetration rate (which causes less irritation) than other molecules in its class. They also knew that it has an anti-bacteriostatic effect that reduces acne and can lighten brown spots, improve the skin texture and appearance and gradually remove the wrinkles. Moreover, it is an antibacterial so it will work well for acne prone and oily skin Now-a-days Mandelic acid is reported to be used in many cosmetic products like facial moisturizers, exfoliating scrubs, acne treatment creams, facial cleansers, skin lightener, scalp treatment against dandruff, as a facial mask, as an oil controller, foundation cream etc.

Dr. Mark Taylor, a world-renowned dermatologist, first introduced Mandelic acid to the NuCelle Company in late 1996. During the course of his clinical trials, Taylor discovered he was able to use the Mandelic acid safely on pigmented and sensitive skin. NuCelle products feature various products containing Mandelic acid. NuCelle provides lifetime skin care solutions for all skin types, and for all ages, through product formulations that ensure the longevity and vitality of the skin. Treatments include cleansers, scrubs, treatments, moisturizers, mattifiers, and sunscreens. Now they have come up with new products called the Mandelic Marine Complex, which combines Mandelic acid with anti-inflammatory Marine Extracts.

Mandelac is another company dealing with Mandelic aicd in their products which are suitable for a host of problems including blemishes, hyperpigmentation, and wrinkles also has an anti-free radical action. There are four Mandelac products- Mandelac Cream, Mandelac Gel, Mandelac Ampoules, Mandelac Scrub, Mandelac Peeling Pack

MaMa Lotion Mandelic & Malic Acid facial lotion is formulated with 10% Mandelic acid and 10% malic acid (an AHA that comes from apples and pears). This unique 20% AHA formulation is much gentler on skin than many products with a lower potency, which means that you get improved results with less chance of irritation.

Dr. Stanley Jacobs, Bay Area facial plastic surgeon, but Jacobs said his own research study of 16 patients showed that it's previously unknown, and best, property is improvement of skin elasticity. Dr. Jacobs envisions other potential applications for use, such as treatment of burn patients and scar tissue. The medical industry widely uses formulations with Mandelic acid for pre- and post-laser surgery treatment to reduce irritation and infection. He is in the process of creating a sunscreen lotion containing SPF 30 and UVA 12+, and a foaming cleanser, both with Mandelic acid. It is said that if you have sensitive skin but still would like the benefits of AHAs, try products with Mandelic acid.

An interesting verse in the Christian Bible - in verse 9 of the first Chapter of the Book of Ecclesiastes - is as true today as when it was first written in the time of King Solomon.

'History merely repeats itself. It has all been done before. Nothing under the sun is truly new'

How typical is this ancient piece of wisdom when one reads in the popular medical press and on the internet about some of the latest drugs being marketed by the technologically advanced western pharmaceutical industry. Instead of rebelling in opposition to nature, we can turn into more in tune with the gifts endowed by nature. The same well being laws that apply to the animal kingdom also apply to man. We have something valuable to relearn from our wild counterparts. By joining palms with nature and embracing the natural we can enhance our well being and enhance our longevity.

REFERENCES

1. Almatawah, Q.A., and Cowan, D.A. (1999). *Enzyme Microb Technol* 25: 718–724.
2. Almatawah, Q.A., Cramp, R., and Cowan, D.A. (1999). *Extremophiles* 3: 283–291.
3. Ambler R P, Auffret A D, Clarke P H, *FEBS Lett.*, (215):285-290,(1987).
4. Anuradha C. Pandey, Annika A. Durve, Manish S. Pathak, and Madhuri Sharon, *Asian J.Exp. Biol.Sci.*2(1):191-200. 2011.
5. Baker R T, Varshavsky A, *J Biol Chem.*,(270): 12065-12074,(1995).
6. Banerjee A, Sharma R, Banerjee U C, *Applied Microbiology and Biotechnology*, (60):33-44, (2002).
7. Bartling D, Seedorf M, Mithoefer A and Weiler E W, *Eur. J. Biochem.*, (205):417-424, (1990).
8. Bartling D, Seedorf M, Schmidt R, and Weiler E, *Proc. Natl. Acad. Sci. USA*, 91:6021-6025 (1994)
9. Baxter, J., and Cummings, S.P.. *Antonie Van Leeuwenhoek* 90: 1–17 (2006).
10. Beloquil A, Domínguez de María P, Golyshin P,and Ferrer M, *Current Opinion in Microbiology*, (11):240–248, (2008).
11. Berger B, Raadt A, Griegl H, Haydern W, Hechtberger P, Klempie N and Faber K, *Pure and Applied Chemistry*, (64):1085-1088, (1992).
12. Bhushan R, and Agarwal C, *Chromatographia*,(68): 1045-1051,(2008)

13. Bigpatents India- Bioconversion of Mandelonitrile to Mandelic acid by microbes. Application 1014/MUM/2006 published 2008-08-29, filed 2006-06-28. International Info. Classification: C12P7/42 .

14. Bork P, Koonin E V, *Protein Sci.*, (**3**):1344-1346, (1994).

15. Brady D, Beeton A, Zeevaart J, Kgaje C, Rantwijk F, Sheldon R, *Applied Microbiology and Biotechnology.*, (**64**): 76-85, (2004).

16. Brenner C, *Current Opinion in Structural Biology*, (**12**): 775-782, (2002).

17. Bruno C M, Fernandes M, Mateo, C, Kiziak C, Chmura A, Wacker J, Rantwijk F, Stolz A, Sheldon R A, *Advanced Synthesis and Catalysis.*,(**348**):2597 – 2603, (2001).

18. Bunch A W, Rehm H J, Reed G, Puhler A, Stadler P, *In Biotechnology*, (**8**):277-324, (1998).

19. Carlos N, Rene T, Clemente A, Paul R, Brown B,*Biochem J.*, doi: 10.1042/BJ20011714.,(2002).

20. Chen J, Zheng R, Zheng Y, and Shen Y, *Adv Biochem Engin/Biotechnol* (**113**): 33-77, (2009)

21. Choi S Y, and Goo Y M, *Arch. Pharmacol. Res.*,(**9**): 45-47, (1986).

22. Cluness M J, Turner E D, Clements E, Brown D T and O'Reilly C,*J. Gen. Microbiol.*,(**139**):1807-1813, (1993).

23. Cole H, Reynolds T R, Lockyer J M, Buck G A, Denson T, Spence J E, Hymes Wolf B, *J Biol Chem.*,(**269**): 6566-6570, (1994).

24. DeSantis, G., Zhu, Z., Greenberg, W.A., Wong, K., Chaplin, J., Hanson, S.R.,. *J Am Chem Soc* **124**: 9024–9025 (2002)

25. DeSantis, G., Wong, K., Farwell, B., Chatman, K., Zhu, Z., Tomlinson, G., *J Am Chem Soc* **125**:11476–11477 (2003)

26. DeSantis G, Zhu Z, Greenberg W A, Wong K, Chaplin J, Hanson S, Farwell B, Nicholson W L, Rand C L, Weiner D P, Robertson D E, and Burk M J, *Applied and Environmental Microbiology.* (70): 2429-2436, (2004).

27. Drauz K, Waldmann H, Roberts S M Eds., *Enzyme Catalysis in Organic Synthesis*; VCH: Weinheim, Germany. 2002

28. Durve A, Pandey A, Pathak M, and Sharon M, *Asian Journal of Experimental Science*, 23(3):533-539, (2009).

29. Faber K, *Pure Appl. Chem.*,(69):1613-1632, (1997).

30. Fong L Y Y, Fidanza V, Zanesi N, Lock L F, Siracusa L D, Mancini R. Siprashvili Z, Ottey M, Martin S E, Dolsky R, Druck T, McCue P A, Croce C M, Huebner K,*Proc Natl Acad Sci USA.*,(97):4742-4747, (2000).

31. Fruchart J, Behr J, and Melnyk O, *Tetrahedron Letters*, 45 (7):1381-1383, (2004).

32. Fukuzaki H, Aiba Y, Yoshida M, Asano M, Kumakura M, *Macromolecular Chemistry and Physics,* 190(10):2407 – 2415, (2003).

33. Garg VK, Sinha S, Sarkar R, *Dermatol Surg.*, 35(1):59-65, (2009)

34. Grsic S, Sauerteig S, Neuhaus K, Albrecht M, Rossiter J, Ludwig-Müller J, *Journal of plant physiology*, (153): 446-456 (1998).

35. Gupta V, Gaind S, Verma P, Sood N and Srivastava A, *African Journal of Microbiology Research*, 4(11): 1148-1153, (2010).

36. Harper D B, *Biochem J*, (165): 309-319, (1977).

37. Harper D B, *J Biochem,* (17)1677 683,(1985).

38. He Y, Xu J, Xu Y, Ouyang L and Pan J, *Chinese Chemical Letters*, 18(6):677-680, (2007).

39. He Y, Zhang Z, Xu J, Liu Y, *J Ind Microbiol Biotechnol* DOI 10.1007/s10295-010-0720-y (2010)

40. Honicke-Schmidt P, Schneider M P, *Journal of Biotechnology*, (**6**): 165-174, (1998).

41. Howden A. J. M. and Preston, G. M, *Microbial Biotechnology*, (**2**): 441–451. (2009).

42. Huang W, Jia J, Cummings J, Nelson M, Schneider G, Lindqvist Y, *Structure.*, (**5**): 691-699, (1997).

43. Jandhyala DM, Willson RC, Sewell BT, Benedik MJ, *Appl Microbiol Biotechnol*, **68**(3): 327-335, (2005)

44. Kabaivanova L, Dobreva E, Dimitrov P and Emanuilova E, *Journal of Industrial Microbiology and Biotechnology*, (**32**): 7-11, (2005).

45. Kakeya H, Sakai N, Sugai T, and Ohta H, *Tetrahedron Lett.*, (**32**):1343-1346, (1991).

46. Kaplan O, Vejvoda V, Charvátová-Pišvejcová A, and Martínková L, *Journal of Industrial Microbiology and Biotechnology*, **33**(11): 891-896, (2006).

47. Kaul P, Banerjee A, Mayilrajb A and Banerjeea U, *Tetrahedron: Asymmetry*, (**15**): 207–211,(2004).

48. Kezic S, Jakasa I, Wenker M, Chromatogr J, **738**: 39, (2000.)

49. Klempier N,Raadt A, Faber K, and Griengl H,*Tetrahedron Lett.*, (**32**): 341-344,(1991).

50. Kobayashi M, Nishiyama M, Nagasawa T, Horinouchi S, Beppu T and Yamada H, *BiochEm. Biophys. Acta*, (**1129**):23-33, (1991).

51. Kobayashi M, Komeda H, Yanaka N, Nagasawa T and Yamada H,*J. Biol. Chem.*, (**267**):20746-20751, (1992).

52. Kobayashi M, Yanaka N, Nagasawa T and Yamada H, *Biochemistry*, (31): 9000-9007, (1992).

53. Kobayashi M, Izui H, Nagasawat H, and Yamada H, *Proc. Nail. Acad. Sci. USA*, (90): 247-251, (1993).

54. Kobayashi, M., and Shimizu, S. Versatile nitrilases – nitrile-hydrolyzing enzymes. *FEMS. Microbiol Lett* 120: 217–223 (1994).

55. Kobayashi, M., Suzuki, T., Fujita, T., Masuda, M., and Shimizu, S. *Proc Natl Acad Sci USA* 92: 714–718 (1995).

56. Kvalnes-Krick K L, Traut T W, *J Biol Chem*, (268): 5686-5693, (1993).

57. Lancaster M, *JNC*, (80):1141, (2003).

58. Langdahl B, Bisp P and Ingvorsen K, *Microbiology*, (142)145-154, (1996).

59. Louwrier A, Knowles C J,*Biotechnol Appl Biochem.*, (25):143-149, (1997).

60. Marcotte E M, Pellegrini M, Rice D W, Yeates T O, and Eisenberg D, *Science.*, (285): 751-753, (1999).

61. Marcotte E, Pellegrini M, Thompson M, Yeates T, Eisenberg D,*Nature.*, (402):83-86, (1999).

62. Martínková L, Klempier N, Preiml M, Ovesná M, Kuzma M, Mylerová V and Kren V, *Can. J. Chem. Rev. Can. Chem,*, (80):724-727, (2002).

63. Martínková L, and Křen V, *Current Opinion in Chemical Biology*, 14(2):130-137, (2010).

64. Mateo C, Fernandes B, Rantwijk F, Stolz A and Sheldon R A, *Journal of Molecular Catalysis B: Enzymatic.*, (38):154-157, (2006).

65. Mathew, C.D., Nagasawa, T., Kobayashi, M., and Yamada, H. Nitrilase-catalyzed production of nicotinic-acid from 3-cyanopyridine in

Rhodococcus rhodochrous J1. *Appl Environ Microbiol* **54**: 1030–1032 (1988).

66. Mathew K, Ravi S, Unny V, Sivaprasad N, *Journal of Radioanalytical and Nuclear Chemistry*, **268**(3):651–652, (2006).

67. Mayaux J F, Cerbelaud E, Soubrier F, Faucher D and Petre D, *J. Bacteriol.*, (**172**):6764-6773, (1990).

68. Michels P C and Rosazza J P N, *Manual of Industrial microbiology and Biotechnology*, 2nd Edition, ASM Press, Washington-DC,(1999).

69. Mukaiyama T, *et al.*, *Chem. Int. Ed. Engl.*, **15**: 94-103, (1976).

70. Mueller, P., Egorova, K., Vorgias, C.E., Boutou, E., Trauthwein, H., Verseck, S., and Antranikian, G. *Protein Expr Purif* **47**: 672–681 (2006).

71. Nagasawa T, Mauger J and Yamada H, Eur. *J. Biochem.*, (**194**):765-771, (1990).

72. Nagasawa T, Wieser M, Nakamura T, Iwahara H, Yoshida T, Gekko K, *Eur J Biochem*,,(**267**):138-44, (2000).

73. Nakai T, Hasegawa T, Yamashita E, Yamamoto M, Kumasaka T, Ueki T, Nanba H, Ikenaka Y, Takahashi S, Sato M, *Structure*,(**8**):729-737, (2000).

74. Nigam V, Khandelwal R, Gothwal R, Mohan M, Choudhury B, Vidyarthi A and Ghosh P, *J. Biosci.* **34**(1) (2009).

75. Nishiyama M, Horinouchi S, Kobayashi M, Nagasawa T, Yamada H and Beppu T, *J. Bacteriol.*, (**173**):2645-2432,(1991).

76. Normanly, J., and Bartel, B.. *Curr Opin Plant Biol* **2**: 207–213 (1999).

77. Normanly, J., Grisafi, P., Fink, G.R., and Bartel, B. *Plant Cell* **9**: 1781–1790 (1997).

78. Novo C, Tata R, Clemente A, and Brown P, *FEBS Lett.*, (**367**):275-279, (1995).

79. O'Reilly, C., and Turner, P.D. The nitrilase family of CN hydrolysing enzymes – a comparative study. *J Appl Microbiol* **95**: 1161–1174 (2003).

80. Osswald, S., Wajant, H., and Effenberger, F.. *Eur J Biochem* **269**: 680–687 (2002).

81. Pace N R, *Science.* (**276**):734-740, (1997).

82. Pace, H.C., Hodawadekar, S.C., Draganescu, A., Huang, J., Bieganowski, P., Pekarsky, Y., et al. *Curr Biol* **10**: 907–917 (2000).

83. Pace, H.C., and Brenner, C. *Genome Biol 2*: REVIEWS0001 (2001).

84. Pathak M, Durve A, Pandey A and Sharon M, *Asian Journal of Chemistry*, **20**(5):3502-3506, (2008).

85. Patricelli M P, Cravatt B F,*J Biol Chem*.19177-19184, (2000).

86. Pekarsky Y, Campiglio M, Siprashvili Z, Druck T, Sedkov Y, Tillib S, Draganescu A, Wermuth P, Rothman J H, Huebner K, Buchberg A M, Mazo A, Brenner C and Croce C M,*Proc Natl Acad Sci USA.*, (**95**):8744-8749,(1998).

87. Piotrowski M, *Phytochemistry*, **69**(15): 2655-2667, (2008).

88. Pollmann S, Neu D, Lehmann T, Berkowitz O, Schafer T, Weiler E W,. DOI- 10.1007/s00425-006-0304-2.,(2002).

89. Qing Z, Ao F, Yuanshan W, Xiaoqin Z, Zhao W, Minghuo W, and Yuguo Z,*Appl. Environ. Microbiol.* Doi-10.1128/AEM.01089-07., (2007).

90. Quirino B, Normanly J and Amasino R M, *Plant Molecular Biology.*, (**40**):267-278, (1999).

91. Raadt A de, Klempier N, Faber K and Griengl H,*J. Chem. Soc.*, (**64**): 1085-1088, (1992).

92. Rappe M S, Giovannoni S J, *Annu Rev Microbiol*,(**57**):369-394, (2000).

93. Rawlings N D, Barrett A J, *Nucleic Acids Res.*, (**28**):323-325, (2000).

94. Rey P, Rossi I, Taillades J, Gros G, and Nore O, *J. Agric. Food Chem.*, **52** (**26**): 8155–8162,(2004).

95. Robertson, D.E., Chaplin, J.A., DeSantis, G., Podar, M., Madden, M., Chi, E., et al. Exploring nitrilase sequence space for enantioselective catalysis. *Appl Environ Microbiol* 70: 2429–2436 (2004).

96. Rustler S, Muller A, Windeisen V, Chmura A, Fernandes B C M, Kiziak C and Stolz A, *Enzyme and Microbial Technology*, (**40**): 598-606, (2007).

97. Scapolla C, Bianchi L, Galatini A,Rocca VM, Benatti U,Damonte G, *GIFC*, (2010)

98. Singh, R., Sharma, R., Tewari, N., and Rawat, D.S. *Nitrilase and its application as a 'green' catalyst.* Chem Biodivers **3**: 1279–1287 (2006).

99. Stalker D M, Malyj L D and Mcbride K E,*J. Biol. Chem.*, (**263**): 6310-6314, (1998).

100. Stevenson D E, Feng R, Storer A C, FEBS Lett., (**277**):112-114, (1990).

101. Thimann K V and Mahadevan S, *Archives of Biochemistry and Biophysics*, (**105**): 133-141, (1964).

102. Undheim K, Tveita PO, *Acta Chem. Scand.*, 25:5-17, (1971).

103. US Patent 7022876 - Preparation of Mandelic acid derivatives (2006).

104. US Patent 5441888 - Process for producing D-Mandelic acid from benzoylformic acid (1995).

105. USPTO Patent Application 20070218136, New Mandelic acid derivatives and their use as thrombin inhibitors.

106. Varala R, Kotra V, Mujahid M, Ramesh N, Ganapaty S, Srinivas R, *Indian journal of chemistry*, **47**(8): 1243-1248, (2008).

107. Walker K, Sjogren, E, Matthews T R; *J. Medicinal Chemistry*, **28**:1673-1679, (1985).

108. Wang P and Van Etten H D, *Biochem. Biophys. Res. Commun*, (**187**):1048-1054, (1992).

109. Wieser M, Nagasawa T, *In Stereoselective Biocatalysis*, (**17**): 461-486, (2000).

110. Williams G, Stickler D, *The Journal of Urology*, **178**(2):697 – 701, (2003).

111. Wong C H, Whitesides G M *Enzymes in Synthetic Organic Chemistry*, 2nd ed.; Pergamon: New York,(1994).

112. Xu H, and Chen Y, *Ultrasonics Sonochemistry*, **15**(6):930-932, (2008).

113. Yamamoto K, Oishi K, Fujimatsu I, Komatsu K,*Applied and Environmental Microbiology*, (**57**):3028- 3032, (1991).

114. Yazbeck D, Durao P, Xie Z, Tao J, *Journal of Molecular Catalysis B: Enzymatic*, (**39**):156–159, (2006).

115. Yoshikawa A, Isono S, Sheback A, Isono K,*Mol Gen Genet.*, (**209**):481-488,(1997).

116. Zalkin H, Smith JL, *Advances in Enzymology and Related Areas of Molecular Biology*, (**72**):87-144, (1998).

117. Zaneveld L, Anderson R, Diao X, Waller D, Chany C, Feathergill K, Doncel G, Cooper M, Herold B, *Fertility and Sterility*,**78**(5):1107-1115,(2002).

118. Zhu Q, Fan A, Wang Y, Zhu X, Wang Z, Wu Z, and Zheng Z, *Applied and Environmental Microbiology*, **73**(19): 6053-6057, (2007).

119. Zhu D, Mukherjee C, Bichl R, Hua L, *Journal of Biotechnology*,(**129**): 645-650, (2007).

120. Zougagh M, Arce L., Ríos A, Valcárcel M, *Journal of Chromatography*, **1104**,:331, (2006)

Information on Biosynthesis, Bioconversion, biotransformation, enzymes, enzyme catalysis and uses of enzymes was taken from online sites- wikipedia.

Mandelic acid and its uses information from:

Betterhealthyskin.com
Wikipedia.com
Mandelic acid skincare.com
Mandelic acid Resource Centre
Material Safety date sheet: DL Mandelic acid
Acne.com
Acne Treatment.com
MakeupAlley.com
SkinCareRx.com
DrugSafetysite.com
MaMalotions.com
DermStores.com

INDEX

R

R (-) Mandelic acid
    Retin A
    Rhodococcus
Rosacea

S

S (+) Mandelic acid,
Saccharomyces
ehhipsoldeus
Scrapies,
Silica,
Specific Gravity
Specific Rotation
Styrene

T

Thalidomide
Thin Layer
Chromatography
Titania
    Toxicology
Tretinoin

U

Urinary tract infections
Uromaline

V

Vaginal contraceptive

W

Wrinkles

Z

Zirconia

www.ingramcontent.com/pod-product-compliance
Lightning Source LLC
Chambersburg PA
CBHW021954170526
45157CB00003B/985